Maths for the Biosciences

Maths for the Biosciences

Compiled from:

Foundation Maths
Fifth Edition
Anthony Croft & Robert Davison

PEARSON

Harlow, England • London • New York • Boston • San Francisco • Toronto • Sydney • Auckland • Singapore • Hong Kong
Tokyo • Seoul • Taipei • New Delhi • Cape Town • Sao Paulo • Mexico City • Madrid • Amsterdam • Munich • Paris • Milan

Pearson Education Limited
Edinburgh Gate
Harlow
Essex CM20 2JE

And associated companies throughout the world

Visit us on the World Wide Web at:
www.pearson.com/uk

© Pearson Education Limited 2013

Compiled from:

Foundation Maths
Fifth Edition
Anthony Croft & Robert Davison
ISBN 978-0-273-72940-2
© Pearson Education Limited 1995, 2003, 2006, 2010

ISBN 978-1-78365-881-7

Printed and bound in Great Britain by Bell & Bain Ltd, Glasgow.

Contents

Foreword

As a scientist it is important to understand how mathematics relates to the study of scientific subjects and the scientific method. One of the most important historical advances in science was the use of mathematics and measurement to give precise accounts of observation and experience. This happened around the fifteenth century. Combining mathematics and what, at this time, was natural philosophy was arguably the birth of the modern scientific method. There was a gradual change at this time from a situation where mathematics was admired for its own sake, to one where mathematics was tool that revealed how things must be; if the calculations worked the theory could be supported more strongly. Copernicus used mathematics to support his argument that it was the Earth that moved around the sun, rather than the other way around. His critics believed in philosophical and physical justifications, but with Copernicus his evidence was purely mathematical; this was a major change in the approach to science.

From the use of mathematics in astronomy came its use in terrestrial subjects such as mechanics, navigation and warfare. Because of this, mathematics began to gain higher social, as well as scientific, status. A mathematical approach to these subjects was gradually replacing the more descriptive Aristotelian natural philosophy. In other words, without mathematics it was harder to support a theory. Galileo was a particularly important historical figure in that he used mathematics to understand the world around him, rejecting much of what Aristotle had taught. For example, Aristotle had taught that force must be continuously be applied to an object to keep it in motion. Through experimentation and mathematics Galileo was able to dispute this. Newton and his contemporaries took many of these early ideas forward and, using mathematics, demonstrated many laws of science that lead to a much greater understanding of the world. For example, Newton's and Leibniz's discovery of calculus which enabled scientists to study moving processes such as the expansion of gases. The work of Newton and his contemporaries was acceptable in its time because of the work of their predecessors who had combined mathematics with natural philosophy and made the study of mathematics intellectually and sociably acceptable.

Following the application of mathematics to science came the application of the experiment, replacing the study of experiences; the replacement of the qualitative with the quantitative. From this came the need to measure, not just for mathematicians, but for natural philosophers too. This, around the time of the seventeenth century, lead to the development of instruments for calculating and for measuring. Science and mathematics became inseparable. Some of the most impressive and historically important experiments, such as those carried out in the nineteenth century by Mendel on peas, were completely based on mathematics and a statistical treatment of data. As a scientist, therefore, it is important to develop an appreciation of the power of mathematics and how crucial it is to a thorough and deep understanding of science.

Front cover picture is of a swan feather, taken by Julia Cameron (www.juliacameron.co.uk)

About this book

This book is designed to help students who have a GCSE in mathematics, or equivalent, progress to understand the mathematics required in a biological science degree.

The first half of this book is basic numeracy, in which it is essential to be proficient as a scientist. Much of it will hopefully be revision. The second half looks at applying the numeracy skills and some basic mathematics to biological situations. A general ability to manipulate numbers and think mathematically will help to understand many biological issues, and help interpret data produced through experimentation. Being able to apply basic numeracy and mathematics to unfamiliar situations is crucial. There are problems to try in each chapter that test understanding of each subject covered.

This book has developed from 'Maths for Biologists', but has added content and a revised order of material covered. The original book 'Maths for Biologists' was compiled by Dr Harriet Jones with the kind help and support of many faculty at the School of Biological Sciences at the University of East Anglia with specific thanks to Dr Helen James, Dr Kay Yeoman, Prof Tony Davy and Dr Richard Bowater for advice and proof reading the material. Also Dr Robert Jenkins, from the Dean of Students' Office, who provided invaluable support and advice in the writing of material additional to that supplied from current text books and Dr Andrew Hemmings who supplied the material for the chapter on calculus.
For the current book particular thanks are due to Dr Andrew Hemmings, Dr Helen James, Dr Kay Yeoman, Dr Robert Jenkins and Danielle Gilroy. The feedback from many cohorts of undergraduate students at UEA who have completed the first year maths module, as well as the many members of faculty who teach small-group maths seminars, has been invaluable in helping to find effective ways of helping students develop confidence in maths as it applies to the biosciences. This feedback has greatly informed the development of this book.

Dr Harriet Jones is a Senior Lecturer in the School of Biological Sciences, University of East Anglia. She researches the student transition from school, and other pre-university environments, to university.
Harriet.jones@uea.ac.uk

Arithmetic of whole numbers 1

Objectives: This chapter:

- explains the rules for adding, subtracting, multiplying and dividing positive and negative numbers
- explains what is meant by an integer
- explains what is meant by a prime number
- explains what is meant by a factor
- explains how to prime factorise an integer
- explains the terms 'highest common factor' and 'lowest common multiple'

1.1 Addition, subtraction, multiplication and division

Arithmetic is the study of numbers and their manipulation. A clear and firm understanding of the rules of arithmetic is essential for tackling everyday calculations. Arithmetic also serves as a springboard for tackling more abstract mathematics such as algebra and calculus.

The calculations in this chapter will involve mainly whole numbers, or **integers** as they are often called. The **positive integers** are the numbers

$$1, 2, 3, 4, 5 \ldots$$

and the **negative integers** are the numbers

$$\ldots -5, -4, -3, -2, -1$$

The dots (...) indicate that this sequence of numbers continues indefinitely. The number 0 is also an integer but is neither positive nor negative.

To find the **sum** of two or more numbers, the numbers are added together. To find the **difference** of two numbers, the second is subtracted from the first. The **product** of two numbers is found by multiplying the numbers together. Finally, the **quotient** of two numbers is found by dividing the first number by the second.

WORKED EXAMPLE

1.1   (a) Find the sum of 3, 6 and 4.
  (b) Find the difference of 6 and 4.
  (c) Find the product of 7 and 2.
  (d) Find the quotient of 20 and 4.

Solution   (a) The sum of 3, 6 and 4 is

$$3 + 6 + 4 = 13$$

  (b) The difference of 6 and 4 is

$$6 - 4 = 2$$

  (c) The product of 7 and 2 is

$$7 \times 2 = 14$$

  (d) The quotient of 20 and 4 is $\frac{20}{4}$, that is 5.

When writing products we sometimes replace the sign $\times$ by '$\cdot$' or even omit it completely. For example, $3 \times 6 \times 9$ could be written as $3 \cdot 6 \cdot 9$ or $(3)(6)(9)$.

On occasions it is necessary to perform calculations involving negative numbers. To understand how these are added and subtracted consider Figure 1.1, which shows a number line.

Figure 1.1
The number line

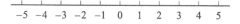

Any number can be represented by a point on the line. Positive numbers are on the right-hand side of the line and negative numbers are on the left. From any given point on the line, we can add a positive number by moving that number of places to the right. For example, to find the sum $5 + 3$, start at the point 5 and move 3 places to the right, to arrive at 8. This is shown in Figure 1.2.

Figure 1.2
To add a positive number, move that number of places to the right

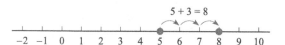

To subtract a positive number, we move that number of places to the left. For example, to find the difference $5 - 7$, start at the point 5 and move 7 places to the left to arrive at -2. Thus $5 - 7 = -2$. This is shown in Figure 1.3. The result of finding $-3 - 4$ is also shown to be -7.

Figure 1.3
To subtract a positive number, move that number of places to the left

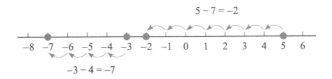

To add or subtract a negative number, the motions just described are reversed. So, to add a negative number, we move to the left. To subtract a negative number we move to the right. The result of finding $2 + (-3)$ is shown in Figure 1.4.

Figure 1.4
Adding a negative number involves moving to the left

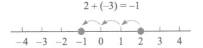

We see that $2 + (-3) = -1$. Note that this is the same as the result of finding $2 - 3$, so that adding a negative number is equivalent to subtracting a positive number.

Figure 1.5
Subtracting a negative number involves moving to the right

The result of finding $5 - (-3)$ is shown in Figure 1.5.
We see that $5 - (-3) = 8$. This is the same as the result of finding $5 + 3$, so subtracting a negative number is equivalent to adding a positive number.

Key point   Adding a negative number is equivalent to subtracting a positive number. Subtracting a negative number is equivalent to adding a positive number.

WORKED EXAMPLE

1.2   Evaluate (a) $8 + (-4)$, (b) $-15 + (-3)$, (c) $-15 - (-4)$.

Solution   (a) $8 + (-4)$ is equivalent to $8 - 4$, that is 4.

(b) Because adding a negative number is equivalent to subtracting a positive number we find $-15 + (-3)$ is equivalent to $-15 - 3$, that is -18.

(c) $-15 - (-4)$ is equivalent to $-15 + 4$, that is -11.

When we need to multiply or divide negative numbers, care must be taken with the **sign** of the answer; that is, whether the result is positive or negative. The following rules apply for determining the sign of the answer when multiplying or dividing positive and negative numbers.

Key point	
$(\text{positive}) \times (\text{positive}) = \text{positive}$ and	$\dfrac{\text{positive}}{\text{positive}} = \text{positive}$
$(\text{positive}) \times (\text{negative}) = \text{negative}$	
$(\text{negative}) \times (\text{positive}) = \text{negative}$	
$(\text{negative}) \times (\text{negative}) = \text{positive}$	$\dfrac{\text{positive}}{\text{negative}} = \text{negative}$
	$\dfrac{\text{negative}}{\text{positive}} = \text{negative}$
	$\dfrac{\text{negative}}{\text{negative}} = \text{positive}$

WORKED EXAMPLE

1.3   Evaluate

(a) $3 \times (-2)$ (b) $(-1) \times 7$ (c) $(-2) \times (-4)$ (d) $\dfrac{12}{(-4)}$ (e) $\dfrac{-8}{4}$ (f) $\dfrac{-6}{-2}$

Solution (a) We have a positive number, 3, multiplied by a negative number, -2, and so the result will be negative:

$$3 \times (-2) = -6$$

(b) $(-1) \times 7 = -7$

(c) Here we have two negative numbers being multiplied and so the result will be positive:

$$(-2) \times (-4) = 8$$

(d) A positive number, 12, divided by a negative number, -4, gives a negative result:

$$\frac{12}{-4} = -3$$

(e) A negative number, -8, divided by a positive number, 4, gives a negative result:

$$\frac{-8}{4} = -2$$

(f) A negative number, -6, divided by a negative number, -2, gives a positive result:

$$\frac{-6}{-2} = 3$$

Self-assessment questions 1.1

1. Explain what is meant by an integer, a positive integer and a negative integer.

2. Explain the terms sum, difference, product and quotient.

3. State the sign of the result obtained after performing the following calculations:
   (a) $(-5) \times (-3)$ (b) $(-4) \times 2$ (c) $\frac{7}{-2}$ (d) $\frac{-8}{-4}$.

Exercise 1.1

1. Without using a calculator, evaluate each of the following:
   (a) $6 + (-3)$ (b) $6 - (-3)$
   (c) $16 + (-5)$ (d) $16 - (-5)$
   (e) $27 - (-3)$ (f) $27 - (-29)$
   (g) $-16 + 3$ (h) $-16 + (-3)$
   (i) $-16 - 3$ (j) $-16 - (-3)$
   (k) $-23 + 52$ (l) $-23 + (-52)$
   (m) $-23 - 52$ (n) $-23 - (-52)$

2. Without using a calculator, evaluate
   (a) $3 \times (-8)$ (b) $(-4) \times 8$ (c) $15 \times (-2)$
   (d) $(-2) \times (-8)$ (e) $14 \times (-3)$

3. Without using a calculator, evaluate
   (a) $\frac{15}{-3}$ (b) $\frac{21}{7}$ (c) $\frac{-21}{7}$ (d) $\frac{-21}{-7}$ (e) $\frac{21}{-7}$
   (f) $\frac{-12}{2}$ (g) $\frac{-12}{-2}$ (h) $\frac{12}{-2}$

4. Find the sum and product of (a) 3 and 6, (b) 10 and 7, (c) 2, 3 and 6.

5. Find the difference and quotient of (a) 18 and 9, (b) 20 and 5, (c) 100 and 20.

1.2 The BODMAS rule

When evaluating numerical expressions we need to know the order in which addition, subtraction, multiplication and division are carried out. As a simple example, consider evaluating $2 + 3 \times 4$. If the addition is carried out first we get $2 + 3 \times 4 = 5 \times 4 = 20$. If the multiplication is carried out first we get $2 + 3 \times 4 = 2 + 12 = 14$. Clearly the order of carrying out numerical operations is important. The BODMAS rule tells us the order in which we must carry out the operations of addition, subtraction, multiplication and division.

Key point	BODMAS stands for	
	Brackets ( )	First priority
	Of $\times$	Second priority
	Division $\div$	Second priority
	Multiplication $\times$	Second priority
	Addition $+$	Third priority
	Subtraction $-$	Third priority

This is the order of carrying out arithmetical operations, with bracketed expressions having highest priority and subtraction and addition having the lowest priority. Note that 'Of', 'Division' and 'Multiplication' have equal priority, as do 'Addition' and 'Subtraction'. 'Of' is used to show multiplication when dealing with fractions: for example, find $\frac{1}{2}$ of 6 means $\frac{1}{2} \times 6$.

   If an expression contains only multiplication and division, we evaluate by working from left to right. Similarly, if an expression contains only addition and subtraction, we also evaluate by working from left to right.

WORKED EXAMPLES

1.4   Evaluate

(a) $2 + 3 \times 4$ (b) $(2 + 3) \times 4$

Solution   (a) Using the BODMAS rule we see that multiplication is carried out first. So

$$2 + 3 \times 4 = 2 + 12 = 14$$

(b) Using the BODMAS rule we see that the bracketed expression takes priority over all else. Hence

$$(2 + 3) \times 4 = 5 \times 4 = 20$$

1.5   Evaluate

(a) $4 - 2 \div 2$ (b) $1 - 3 + 2 \times 2$

Solution   (a)   Division is carried out before subtraction, and so

$$4 - 2 \div 2 = 4 - \frac{2}{2} = 3$$

(b)   Multiplication is carried out before subtraction or addition:
$$1 - 3 + 2 \times 2 = 1 - 3 + 4 = 2$$

1.6   Evaluate

(a) $(12 \div 4) \times 3$ (b) $12 \div (4 \times 3)$

Solution   Recall that bracketed expressions are evaluated first.

(a) $(12 \div 4) \times 3 = \left(\frac{12}{4}\right) \times 3 = 3 \times 3 = 9$

(b) $12 \div (4 \times 3) = 12 \div 12 = 1$

Example 1.6 shows the importance of the position of brackets in an expression.

Self-assessment questions 1.2

1.   State the BODMAS rule used to evaluate expressions.

2.   The position of brackets in an expression is unimportant. True or false?

Exercise 1.2

MyMathLab

1.   Evaluate the following expressions:
(a) $6 - 2 \times 2$ (b) $(6 - 2) \times 2$
(c) $6 \div 2 - 2$ (d) $(6 \div 2) - 2$
(e) $6 - 2 + 3 \times 2$ (f) $6 - (2 + 3) \times 2$
(g) $(6 - 2) + 3 \times 2$ (h) $\frac{16}{-2}$ (i) $\frac{-24}{-3}$
(j) $(-6) \times (-2)$ (k) $(-2)(-3)(-4)$

2.   Place brackets in the following expressions to make them correct:
(a) $6 \times 12 - 3 + 1 = 55$
(b) $6 \times 12 - 3 + 1 = 68$
(c) $6 \times 12 - 3 + 1 = 60$
(d) $5 \times 4 - 3 + 2 = 7$
(e) $5 \times 4 - 3 + 2 = 15$
(f) $5 \times 4 - 3 + 2 = -5$

1.3 Prime numbers and factorisation

A **prime number** is a positive integer, larger than 1, which cannot be expressed as the product of two smaller positive integers. To put it another way, a prime number is one that can be divided exactly only by 1 and itself.

For example, $6 = 2 \times 3$, so 6 can be expressed as a product of smaller numbers and hence 6 is not a prime number. However, 7 is prime. Examples of prime numbers are 2, 3, 5, 7, 11, 13, 17, 19, 23. Note that 2 is the only even prime.

Factorise means 'write as a product'. By writing 12 as 3×4 we have factorised 12. We say 3 is a **factor** of 12 and 4 is also a factor of 12. The way in which a number is factorised is not unique: for example, 12 may be expressed as 3×4 or 2×6. Note that 2 and 6 are also factors of 12.

When a number is written as a product of prime numbers we say the number has been **prime factorised**.

To prime factorise a number, consider the technique used in the following examples.

WORKED EXAMPLES

1.7   Prime factorise the following numbers:

(a) 12   (b) 42   (c) 40   (d) 70

Solution   (a) We begin with 2 and see whether this is a factor of 12. Clearly it is, so we write

$$12 = 2 \times 6$$

Now we consider 6. Again 2 is a factor so we write

$$12 = 2 \times 2 \times 3$$

All the factors are now prime, that is the prime factorisation of 12 is $2 \times 2 \times 3$.

(b) We begin with 2 and see whether this is a factor of 42. Clearly it is and so we can write

$$42 = 2 \times 21$$

Now we consider 21. Now 2 is not a factor of 21, so we examine the next prime, 3. Clearly 3 is a factor of 21 and so we can write

$$42 = 2 \times 3 \times 7$$

All the factors are now prime, and so the prime factorisation of 42 is $2 \times 3 \times 7$.

(c) Clearly 2 is a factor of 40,

$$40 = 2 \times 20$$

Clearly 2 is a factor of 20,

$$40 = 2 \times 2 \times 10$$

Again 2 is a factor of 10,

$$40 = 2 \times 2 \times 2 \times 5$$

All the factors are now prime. The prime factorisation of 40 is $2 \times 2 \times 2 \times 5$.

(d) Clearly 2 is a factor of 70,

$$70 = 2 \times 35$$

We consider 35: 2 is not a factor, 3 is not a factor, but 5 is:

$$70 = 2 \times 5 \times 7$$

All the factors are prime. The prime factorisation of 70 is $2 \times 5 \times 7$.

1.8   Prime factorise 2299.

Solution   We note that 2 is not a factor and so we try 3. Again 3 is not a factor and so we try 5. This process continues until we find the first prime factor. It is 11:

$$2299 = 11 \times 209$$

We now consider 209. The first prime factor is 11:

$$2299 = 11 \times 11 \times 19$$

All the factors are prime. The prime factorisation of 2299 is $11 \times 11 \times 19$.

Self-assessment questions 1.3

1. Explain what is meant by a prime number.

2. List the first 10 prime numbers.

3. Explain why all even numbers other than 2 cannot be prime.

Exercise 1.3

1. State which of the following numbers are prime numbers:
   (a) 13   (b) 1000   (c) 2   (c) 29   (d) $\frac{1}{2}$

2. Prime factorise the following numbers:
   (a) 26   (b) 100   (c) 27   (d) 71   (e) 64   (f) 87   (g) 437   (h) 899

3. Prime factorise the two numbers 30 and 42. List any prime factors which are common to both numbers.

1.4 Highest common factor and lowest common multiple

Highest common factor

Suppose we prime factorise 12. This gives $12 = 2 \times 2 \times 3$. From this prime factorisation we can deduce all the factors of 12:

2 is a factor of 12
3 is a factor of 12
$2 \times 2 = 4$ is a factor of 12
$2 \times 3 = 6$ is a factor of 12

Hence 12 has factors 2, 3, 4 and 6, in addition to the obvious factors of 1 and 12.

Similarly we could prime factorise 18 to obtain $18 = 2 \times 3 \times 3$. From this we can list the factors of 18:

2 is a factor of 18
3 is a factor of 18
$2 \times 3 = 6$ is a factor of 18
$3 \times 3 = 9$ is a factor of 18

The factors of 18 are 1, 2, 3, 6, 9 and 18. Some factors are common to both 12 and 18. These are 2, 3 and 6. These are **common factors** of 12 and 18. The highest common factor of 12 and 18 is 6.

The highest common factor of 12 and 18 can be obtained directly from their prime factorisation. We simply note all the primes common to both factorisations:

$$12 = 2 \times 2 \times 3 \qquad 18 = 2 \times 3 \times 3$$

Common to both is 2×3. Thus the highest common factor is $2 \times 3 = 6$. Thus 6 is the highest number that divides exactly into both 12 and 18.

Key point

> Given two or more numbers the **highest common factor** (h.c.f.) is the largest (highest) number that is a factor of all the given numbers.
> The highest common factor is also referred to as the **greatest common divisor** (g.c.d).

WORKED EXAMPLES

1.9   Find the h.c.f. of 12 and 27.

Solution   We prime factorise 12 and 27:

$$12 = 2 \times 2 \times 3 \qquad 27 = 3 \times 3 \times 3$$

Common to both is 3. Thus 3 is the h.c.f. of 12 and 27. This means that 3 is the highest number that divides both 12 and 27.

1.10   Find the h.c.f. of 28 and 210.

Solution   The numbers are prime factorised:

$$28 = 2 \times 2 \times 7$$
$$210 = 2 \times 3 \times 5 \times 7$$

The factors that are common are identified: a 2 is common to both and a 7 is common to both. Hence both numbers are divisible by $2 \times 7 = 14$. Since this number contains all the common factors it is the highest common factor.

1.11   Find the h.c.f. of 90 and 108.

Solution   The numbers are prime factorised:

$$90 = 2 \times 3 \times 3 \times 5$$
$$108 = 2 \times 2 \times 3 \times 3 \times 3$$

The common factors are 2, 3 and 3 and so the h.c.f. is $2 \times 3 \times 3$, that is 18. This is the highest number that divides both 90 and 108.

1.12   Find the h.c.f. of 12, 18 and 20.

Solution   Prime factorisation yields

$$12 = 2 \times 2 \times 3 \qquad 18 = 2 \times 3 \times 3 \qquad 20 = 2 \times 2 \times 5$$

There is only one factor common to all three numbers: it is 2. Hence 2 is the h.c.f. of 12, 18 and 20.

Lowest common multiple

Suppose we are given two or more numbers and wish to find numbers into which all the given numbers will divide. For example, given 4 and 6 we see that they both divide exactly into 12, 24, 36, 48, 60 and so on. The smallest number into which they both divide is 12. We say 12 is the **lowest common multiple** of 4 and 6.

Key point   The lowest common multiple (l.c.m.) of a set of numbers is the smallest (lowest) number into which all the given numbers will divide exactly.

WORKED EXAMPLE

1.13   Find the l.c.m. of 6 and 10.

Solution   We seek the smallest number into which both 6 and 10 will divide exactly. There are many numbers into which 6 and 10 will divide, for example 60,

120, 600, but we are seeking the smallest such number. By inspection, the smallest such number is 30. Thus the l.c.m. of 6 and 10 is 30.

A more systematic method of finding the l.c.m. involves the use of prime factorisation.

WORKED EXAMPLES

1.14   Find the l.c.m. of 15 and 20.

Solution   As a first step, the numbers are prime factorised:

$$15 = 3 \times 5 \qquad 20 = 2 \times 2 \times 5$$

Since 15 must divide into the l.c.m., then the l.c.m. must contain the factors of 15, that is 3×5. Similarly, as 20 must divide into the l.c.m., then the l.c.m. must also contain the factors of 20, that is $2 \times 2 \times 5$. The l.c.m. is the smallest number that contains both of these sets of factors. Note that the l.c.m. will contain only 2s, 3s and 5s as its prime factors. We now need to determine how many of these particular factors are needed.

To determine the l.c.m. we ask 'How many factors of 2 are required?', 'How many factors of 3 are required?', 'How many factors of 5 are required?'

The highest number of 2s occurs in the factorisation of 20. Hence the l.c.m. requires two factors of 2. Consider the number of 3s required. The highest number of 3s occurs in the factorisation of 15. Hence the l.c.m. requires one factor of 3. Consider the number of 5s required. The highest number of 5s is 1 and so the l.c.m. requires one factor of 5. Hence the l.c.m. is $2 \times 2 \times 3 \times 5 = 60$.

Hence 60 is the smallest number into which both 15 and 20 will divide exactly.

1.15   Find the l.c.m. of 20, 24 and 25.

Solution   The numbers are prime factorised:

$$20 = 2 \times 2 \times 5 \qquad 24 = 2 \times 2 \times 2 \times 3 \qquad 25 = 5 \times 5$$

By considering the prime factorisations of 20, 24 and 25 we see that the only primes involved are 2, 3 and 5. Hence the l.c.m. will contain only 2s, 3s and 5s.

Consider the number of 2s required. The highest number of 2s required is three from factorising 24. The highest number of 3s required is one, again from factorising 24. The highest number of 5s required is two, found from factorising 25. Hence the l.c.m. is given by

$$\text{l.c.m.} = 2 \times 2 \times 2 \times 3 \times 5 \times 5 = 600$$

Hence 600 is the smallest number into which 20, 24 and 25 will all divide exactly.

Self-assessment questions 1.4

1. Explain what is meant by the h.c.f. of a set of numbers.

2. Explain what is meant by the l.c.m. of a set of numbers.

Exercise 1.4

1. Calculate the h.c.f. of the following sets of numbers:
   (a) 12, 15, 21 (b) 16, 24, 40 (c) 28, 70, 120, 160 (d) 35, 38, 42 (e) 96, 120, 144

2. Calculate the l.c.m. of the following sets of numbers:
   (a) 5, 6, 8 (b) 20, 30 (c) 7, 9, 12 (d) 100, 150, 235 (e) 96, 120, 144

Test and assignment exercises 1

1. Evaluate
   (a) $6 \div 2 + 1$
   (b) $6 \div (2 + 1)$
   (c) $12 + 4 \div 4$
   (d) $(12 + 4) \div 4$
   (e) $3 \times 2 + 1$
   (f) $3 \times (2 + 1)$
   (g) $6 - 2 + 4 \div 2$
   (h) $(6 - 2 + 4) \div 2$
   (i) $6 - (2 + 4 \div 2)$
   (j) $6 - (2 + 4) \div 2$
   (k) $2 \times 4 - 1$
   (l) $2 \times (4 - 1)$
   (m) $2 \times 6 \div (3 - 1)$
   (n) $2 \times (6 \div 3) - 1$
   (o) $2 \times (6 \div 3 - 1)$

2. Prime factorise (a) 56, (b) 39, (c) 74.

3. Find the h.c.f. of
   (a) 8, 12, 14 (b) 18, 42, 66 (c) 20, 24, 30 (d) 16, 24, 32, 160

4. Find the l.c.m. of
   (a) 10, 15 (b) 11, 13 (c) 8, 14, 16 (d) 15, 24, 30

Fractions

2

This chapter:

- explains what is meant by a fraction
- defines the terms 'improper fraction', 'proper fraction' and 'mixed fraction'
- explains how to write fractions in different but equivalent forms
- explains how to simplify fractions by cancelling common factors
- explains how to add, subtract, multiply and divide fractions

2.1 Introduction

The arithmetic of fractions is very important groundwork that must be mastered before topics in algebra such as formulae and equations can be understood. The same techniques that are used to manipulate fractions are used in these more advanced topics. You should use this chapter to ensure that you are confident at handling fractions before moving on to algebra. In all the examples and exercises it is important that you should carry out the calculations without the use of a calculator.

Fractions are numbers such as $\frac{1}{2}$, $\frac{3}{4}$, $\frac{11}{8}$ and so on. In general a fraction is a number of the form $\frac{p}{q}$, where the letters p and q represent whole numbers or integers. The integer q can never be zero because it is never possible to divide by zero.

In any fraction $\frac{p}{q}$ the number p is called the **numerator** and the number q is called the **denominator**.

Key point

$$\text{fraction} = \frac{\text{numerator}}{\text{denominator}} = \frac{p}{q}$$

Suppose that p and q are both positive numbers. If p is less than q, the fraction is said to be a **proper fraction**. So $\frac{1}{2}$ and $\frac{3}{4}$ are proper fractions since the numerator is less than the denominator. If p is greater than or equal to q, the fraction is said to be **improper**. So $\frac{11}{8}$, $\frac{7}{4}$ and $\frac{3}{3}$ are all improper fractions.

If either of p or q is negative, we simply ignore the negative sign when determining whether the fraction is proper or improper. So $-\frac{3}{5}$, $\frac{-7}{21}$ and $\frac{4}{-21}$ are proper fractions, but $\frac{3}{-3}$, $\frac{-8}{2}$ and $-\frac{11}{2}$ are improper.

Note that all proper fractions have a value less than 1.

The denominator of a fraction can take the value 1, as in $\frac{3}{1}$ and $\frac{7}{1}$. In these cases the result is a whole number, 3 and 7.

Self-assessment questions 2.1

1. Explain the terms (a) fraction, (b) improper fraction, (c) proper fraction. In each case give an example of your own.

2. Explain the terms (a) numerator, (b) denominator.

Exercise 2.1

1. Classify each of the following as proper or improper:
   (a) $\frac{9}{17}$ (b) $\frac{-9}{17}$ (c) $\frac{8}{8}$ (d) $-\frac{7}{8}$ (e) $\frac{110}{77}$

2.2 Expressing a fraction in equivalent forms

Given a fraction, we may be able to express it in a different form. For example, you will know that $\frac{1}{2}$ is equivalent to $\frac{2}{4}$. Note that multiplying both numerator and denominator by the same number leaves the value of the fraction unchanged. So, for example,

$$\frac{1}{2} = \frac{1 \times 2}{2 \times 2} = \frac{2}{4}$$

We say that $\frac{1}{2}$ and $\frac{2}{4}$ are **equivalent fractions**. Although they might look different, they have the same value.

Similarly, given the fraction $\frac{8}{12}$ we can divide both numerator and denominator by 4 to obtain

$$\frac{8}{12} = \frac{8/4}{12/4} = \frac{2}{3}$$

so $\frac{8}{12}$ and $\frac{2}{3}$ have the same value and are equivalent fractions.

| Key point | Multiplying or dividing both numerator and denominator of a fraction by the same number produces a fraction having the same value, called an equivalent fraction. |

A fraction is in its **simplest form** when there are no factors common to both numerator and denominator. For example, $\frac{5}{12}$ is in its simplest form, but $\frac{3}{6}$ is not since 3 is a factor common to both numerator and denominator. Its simplest form is the equivalent fraction $\frac{1}{2}$.

To express a fraction in its simplest form we look for factors that are common to both the numerator and denominator. This is done by prime factorising both of these. Dividing both the numerator and denominator by any common factors removes them but leaves an equivalent fraction. This is equivalent to cancelling any common factors. For example, to simplify $\frac{4}{6}$ we prime factorise to produce

$$\frac{4}{6} = \frac{2 \times 2}{2 \times 3}$$

Dividing both numerator and denominator by 2 leaves $\frac{2}{3}$. This is equivalent to cancelling the common factor of 2.

WORKED EXAMPLES

2.1   Express $\frac{24}{36}$ in its simplest form.

Solution   We seek factors common to both numerator and denominator. To do this we prime factorise 24 and 36:

Prime factorisation has been described in §1.3.

$$24 = 2 \times 2 \times 2 \times 3 \qquad 36 = 2 \times 2 \times 3 \times 3$$

The factors $2 \times 2 \times 3$ are common to both 24 and 36 and so these may be cancelled. Note that only common factors may be cancelled when simplifying a fraction. Hence

Finding the highest common factor (h.c.f.) of two numbers is detailed in §1.4.

$$\frac{24}{36} = \frac{\cancel{2} \times \cancel{2} \times 2 \times \cancel{3}}{\cancel{2} \times \cancel{2} \times \cancel{3} \times 3} = \frac{2}{3}$$

In its simplest form $\frac{24}{36}$ is $\frac{2}{3}$. In effect we have divided 24 and 36 by 12, which is their h.c.f.

2.2   Express $\frac{49}{21}$ in its simplest form.

Solution   Prime factorising 49 and 21 gives

$$49 = 7 \times 7 \qquad 21 = 3 \times 7$$

Their h.c.f. is 7. Dividing 49 and 21 by 7 gives

$$\frac{49}{21} = \frac{7}{3}$$

Hence the simplest form of $\frac{49}{21}$ is $\frac{7}{3}$.

Before we can start to add and subtract fractions it is necessary to be able to convert fractions into a variety of equivalent forms. Work through the following examples.

WORKED EXAMPLES

2.3   Express $\frac{3}{4}$ as an equivalent fraction having a denominator of 20.

Solution   To achieve a denominator of 20, the existing denominator must be multiplied by 5. To produce an equivalent fraction both numerator and denominator must be multiplied by 5, so

$$\frac{3}{4} = \frac{3 \times 5}{4 \times 5} = \frac{15}{20}$$

2.4   Express 7 as an equivalent fraction with a denominator of 3.

Solution   Note that 7 is the same as the fraction $\frac{7}{1}$. To achieve a denominator of 3, the existing denominator must be multiplied by 3. To produce an equivalent fraction both numerator and denominator must be multiplied by 3, so

$$7 = \frac{7}{1} = \frac{7 \times 3}{1 \times 3} = \frac{21}{3}$$

Self-assessment questions 2.2

1. All integers can be thought of as fractions. True or false?

2. Explain the use of h.c.f. in the simplification of fractions.

3. Give an example of three fractions that are equivalent.

Exercise 2.2

1. Express the following fractions in their simplest form:

    (a) $\frac{18}{27}$ (b) $\frac{12}{20}$ (c) $\frac{15}{45}$ (d) $\frac{25}{80}$ (e) $\frac{15}{60}$

    (f) $\frac{90}{200}$ (g) $\frac{15}{20}$ (h) $\frac{2}{18}$ (i) $\frac{16}{24}$ (j) $\frac{30}{65}$

    (k) $\frac{12}{21}$ (l) $\frac{100}{45}$ (m) $\frac{6}{9}$ (n) $\frac{12}{16}$ (o) $\frac{13}{42}$

    (p) $\frac{13}{39}$ (q) $\frac{11}{33}$ (r) $\frac{14}{30}$ (s) $-\frac{12}{16}$ (t) $\frac{11}{-33}$

    (u) $\frac{-14}{-30}$

2. Express $\frac{3}{4}$ as an equivalent fraction having a denominator of 28.

3. Express 4 as an equivalent fraction with a denominator of 5.

4. Express $\frac{5}{12}$ as an equivalent fraction having a denominator of 36.

5. Express 2 as an equivalent fraction with a denominator of 4.

6. Express 6 as an equivalent fraction with a denominator of 3.

7. Express each of the fractions $\frac{2}{3}$, $\frac{5}{4}$ and $\frac{5}{6}$ as an equivalent fraction with a denominator of 12.

8. Express each of the fractions $\frac{4}{9}$, $\frac{1}{2}$ and $\frac{5}{6}$ as an equivalent fraction with a denominator of 18.

9. Express each of the following numbers as an equivalent fraction with a denominator of 12:

    (a) $\frac{1}{2}$ (b) $\frac{3}{4}$ (c) $\frac{5}{2}$ (d) 5 (e) 4 (f) 12

2.3 Addition and subtraction of fractions

To add and subtract fractions we first rewrite each fraction so that they all have the same denominator. This is known as the **common denominator**. The denominator is chosen to be the lowest common multiple of the original denominators. Then the numerators only are added or subtracted as appropriate, and the result is divided by the common denominator.

WORKED EXAMPLES

2.5 Find $\frac{2}{3} + \frac{5}{4}$.

Solution The denominators are 3 and 4. The l.c.m. of 3 and 4 is 12. We need to express both fractions with a denominator of 12.

Finding the lowest common multiple (l.c.m.) is detailed in §1.4.

To express $\frac{2}{3}$ with a denominator of 12 we multiply both numerator and denominator by 4. Hence $\frac{2}{3}$ is the same as $\frac{8}{12}$. To express $\frac{5}{4}$ with a denominator of 12 we multiply both numerator and denominator by 3. Hence $\frac{5}{4}$ is the same as $\frac{15}{12}$. So

$$\frac{2}{3} + \frac{5}{4} = \frac{8}{12} + \frac{15}{12} = \frac{8+15}{12} = \frac{23}{12}$$

2.6   Find $\frac{4}{9} - \frac{1}{2} + \frac{5}{6}$.

Solution   The denominators are 9, 2 and 6. Their l.c.m. is 18. Each fraction is expressed with 18 as the denominator:

$$\frac{4}{9} = \frac{8}{18} \qquad \frac{1}{2} = \frac{9}{18} \qquad \frac{5}{6} = \frac{15}{18}$$

Then

$$\frac{4}{9} - \frac{1}{2} + \frac{5}{6} = \frac{8}{18} - \frac{9}{18} + \frac{15}{18} = \frac{8 - 9 + 15}{18} = \frac{14}{18}$$

The fraction $\frac{14}{18}$ can be simplified to $\frac{7}{9}$. Hence

$$\frac{4}{9} - \frac{1}{2} + \frac{5}{6} = \frac{7}{9}$$

2.7   Find $\frac{1}{4} - \frac{5}{9}$.

Solution   The l.c.m. of 4 and 9 is 36. Each fraction is expressed with a denominator of 36. Thus

$$\frac{1}{4} = \frac{9}{36} \qquad \text{and} \qquad \frac{5}{9} = \frac{20}{36}$$

Then

$$\frac{1}{4} - \frac{5}{9} = \frac{9}{36} - \frac{20}{36}$$
$$= \frac{9 - 20}{36}$$
$$= \frac{-11}{36}$$
$$= -\frac{11}{36}$$

Consider the number $2\frac{3}{4}$. This is referred to as a **mixed fraction** because it contains a whole number part, 2, and a fractional part, $\frac{3}{4}$. We can convert this mixed fraction into an improper fraction as follows. Recognise that 2 is equivalent to $\frac{8}{4}$, and so $2\frac{3}{4}$ is $\frac{8}{4} + \frac{3}{4} = \frac{11}{4}$.

The reverse of this process is to convert an improper fraction into a mixed fraction. Consider the improper fraction $\frac{11}{4}$. Now 4 divides into 11 twice leaving a remainder of 3; so $\frac{11}{4} = 2$ remainder 3, which we write as $2\frac{3}{4}$.

WORKED EXAMPLE

2.8   (a) Express $4\frac{2}{5}$ as an improper fraction.

(b) Find $4\frac{2}{5} + \frac{1}{3}$.

Solution   (a) $4\frac{2}{5}$ is a mixed fraction. Note that $4\frac{2}{5}$ is equal to $4 + \frac{2}{5}$. We can write 4 as the equivalent fraction $\frac{20}{5}$. Therefore

$$4\,\frac{2}{5} = \frac{20}{5} + \frac{2}{5}$$

$$= \frac{22}{5}$$

(b) $4\,\frac{2}{5} + \frac{1}{3} = \frac{22}{5} + \frac{1}{3}$

$$= \frac{66}{15} + \frac{5}{15}$$

$$= \frac{71}{15}$$

Self-assessment question 2.3

1. Explain the use of l.c.m. when adding and subtracting fractions.

Exercise 2.3

1. Find
   (a) $\frac{1}{4} + \frac{2}{3}$ (b) $\frac{3}{5} + \frac{5}{3}$ (c) $\frac{12}{14} - \frac{2}{7}$

   (d) $\frac{3}{7} - \frac{1}{2} + \frac{2}{21}$ (e) $1\frac{1}{2} + \frac{4}{9}$

   (f) $2\frac{1}{4} - 1\frac{1}{3} + \frac{1}{2}$ (g) $\frac{10}{15} - 1\frac{2}{5} + \frac{8}{3}$

   (h) $\frac{9}{10} - \frac{7}{16} + \frac{1}{2} - \frac{2}{5}$

2. Find
   (a) $\frac{7}{8} + \frac{1}{3}$ (b) $\frac{1}{2} - \frac{3}{4}$ (c) $\frac{3}{5} + \frac{2}{3} + \frac{1}{2}$

   (d) $\frac{3}{8} + \frac{1}{3} + \frac{1}{4}$ (e) $\frac{2}{3} - \frac{4}{7}$

   (f) $\frac{1}{11} - \frac{1}{2}$ (g) $\frac{3}{11} - \frac{5}{8}$

3. Express as improper fractions:

    (a) $2\frac{1}{2}$ (b) $3\frac{2}{3}$ (c) $10\frac{1}{4}$ (d) $5\frac{2}{7}$

    (e) $6\frac{2}{9}$ (f) $11\frac{1}{3}$ (g) $15\frac{1}{2}$ (h) $13\frac{3}{4}$

    (i) $12\frac{1}{11}$ (j) $13\frac{2}{3}$ (k) $56\frac{1}{2}$

4. Without using a calculator express these improper fractions as mixed fractions:

    (a) $\frac{10}{3}$ (b) $\frac{7}{2}$ (c) $\frac{15}{4}$ (d) $\frac{25}{6}$

2.4 Multiplication of fractions

The product of two or more fractions is found by multiplying their numerators to form a new numerator, and then multiplying their denominators to form a new denominator.

WORKED EXAMPLES

2.9 Find $\frac{4}{9} \times \frac{3}{8}$.

Solution The numerators are multiplied: $4 \times 3 = 12$. The denominators are multiplied: $9 \times 8 = 72$. Hence

$$\frac{4}{9} \times \frac{3}{8} = \frac{12}{72}$$

This may now be expressed in its simplest form:

$$\frac{12}{72} = \frac{1}{6}$$

Hence

$$\frac{4}{9} \times \frac{3}{8} = \frac{1}{6}$$

An alternative, but equivalent, method is to cancel any factors common to both numerator and denominator at the outset:

$$\frac{4}{9} \times \frac{3}{8} = \frac{4 \times 3}{9 \times 8}$$

A factor of 4 is common to the 4 and the 8. Hence

$$\frac{4 \times 3}{9 \times 8} = \frac{1 \times 3}{9 \times 2}$$

A factor of 3 is common to the 3 and the 9. Hence

$$\frac{1 \times 3}{9 \times 2} = \frac{1 \times 1}{3 \times 2} = \frac{1}{6}$$

2.10   Find $\frac{12}{25} \times \frac{2}{7} \times \frac{10}{9}$.

Solution   We cancel factors common to both numerator and denominator. A factor of 5 is common to 10 and 25. Cancelling this gives

$$\frac{12}{25} \times \frac{2}{7} \times \frac{10}{9} = \frac{12}{5} \times \frac{2}{7} \times \frac{2}{9}$$

A factor of 3 is common to 12 and 9. Cancelling this gives

$$\frac{12}{5} \times \frac{2}{7} \times \frac{2}{9} = \frac{4}{5} \times \frac{2}{7} \times \frac{2}{3}$$

There are no more common factors. Hence

$$\frac{12}{25} \times \frac{2}{7} \times \frac{10}{9} = \frac{4}{5} \times \frac{2}{7} \times \frac{2}{3} = \frac{16}{105}$$

2.11   Find $\frac{3}{4}$ of $\frac{5}{9}$.

Recall that 'of' means multiply.

Solution $\frac{3}{4}$ of $\frac{5}{9}$ is the same as $\frac{3}{4} \times \frac{5}{9}$. Cancelling a factor of 3 from numerator and denominator gives $\frac{1}{4} \times \frac{5}{3}$, that is $\frac{5}{12}$. Hence $\frac{3}{4}$ of $\frac{5}{9}$ is $\frac{5}{12}$.

2.12   Find $\frac{5}{6}$ of 70.

Solution   We can write 70 as $\frac{70}{1}$. So

$$\frac{5}{6} \text{ of } 70 = \frac{5}{6} \times \frac{70}{1} = \frac{5}{3} \times \frac{35}{1} = \frac{175}{3} = 58\frac{1}{3}$$

2.13   Find $2\frac{7}{8} \times \frac{2}{3}$.

Solution   In this example the first fraction is a mixed fraction. We convert it to an improper fraction before performing the multiplication. Note that $2\frac{7}{8} = \frac{23}{8}$. Then

$$\frac{23}{8} \times \frac{2}{3} = \frac{23}{4} \times \frac{1}{3}$$

$$= \frac{23}{12}$$

$$= 1\frac{11}{12}$$

Self-assessment question 2.4

1. Describe how to multiply fractions together.

Exercise 2.4

1. Evaluate
   (a) $\frac{2}{3} \times \frac{6}{7}$ (b) $\frac{8}{15} \times \frac{25}{32}$ (c) $\frac{1}{4} \times \frac{8}{9}$
   (d) $\frac{16}{17} \times \frac{34}{48}$ (e) $2 \times \frac{3}{5} \times \frac{5}{12}$
   (f) $2\frac{1}{3} \times 1\frac{1}{4}$ (g) $1\frac{3}{4} \times 2\frac{1}{2}$
   (h) $\frac{3}{4} \times 1\frac{1}{2} \times 3\frac{1}{2}$

2. Evaluate
   (a) $\frac{2}{3}$ of $\frac{3}{4}$ (b) $\frac{4}{7}$ of $\frac{21}{30}$
   (c) $\frac{9}{10}$ of 80 (d) $\frac{6}{7}$ of 42

3. Is $\frac{3}{4}$ of $\frac{12}{15}$ the same as $\frac{12}{15}$ of $\frac{3}{4}$?

4. Find
   (a) $-\frac{1}{3} \times \frac{5}{7}$ (b) $\frac{3}{4} \times -\frac{1}{2}$
   (c) $\left(-\frac{5}{8}\right) \times \frac{8}{11}$ (d) $\left(-\frac{2}{3}\right) \times \left(-\frac{15}{7}\right)$

5. Find
   (a) $5\frac{1}{2} \times \frac{1}{2}$ (b) $3\frac{3}{4} \times \frac{1}{3}$
   (c) $\frac{2}{3} \times 5\frac{1}{9}$ (d) $\frac{3}{4} \times 11\frac{1}{2}$

6. Find
   (a) $\frac{3}{5}$ of $11\frac{1}{4}$ (b) $\frac{2}{3}$ of $15\frac{1}{2}$
   (c) $\frac{1}{4}$ of $-8\frac{1}{3}$

2.5 Division by a fraction

To divide one fraction by another fraction, we invert the second fraction and then multiply. When we invert a fraction we interchange the numerator and denominator.

WORKED EXAMPLES

2.14 Find $\frac{6}{25} \div \frac{2}{5}$.

Solution We invert $\frac{2}{5}$ to obtain $\frac{5}{2}$. Multiplication is then performed. So

$$\frac{6}{25} \div \frac{2}{5} = \frac{6}{25} \times \frac{5}{2} = \frac{3}{25} \times \frac{5}{1} = \frac{3}{5} \times \frac{1}{1} = \frac{3}{5}$$

2.15 Evaluate (a) $1\frac{1}{3} \div \frac{8}{3}$, (b) $\frac{20}{21} \div \frac{5}{7}$.

Solution (a) First we express $1\frac{1}{3}$ as an improper fraction:

$$1\frac{1}{3} = 1 + \frac{1}{3} = \frac{3}{3} + \frac{1}{3} = \frac{4}{3}$$

So we calculate

$$\frac{4}{3} \div \frac{8}{3} = \frac{4}{3} \times \frac{3}{8} = \frac{4}{8} = \frac{1}{2}$$

Hence

$$1\frac{1}{3} \div \frac{8}{3} = \frac{1}{2}$$

(b) $\dfrac{20}{21} \div \dfrac{5}{7} = \dfrac{20}{21} \times \dfrac{7}{5} = \dfrac{4}{21} \times \dfrac{7}{1} = \dfrac{4}{3}$

Self-assessment question 2.5

1. Explain the process of division by a fraction.

Exercise 2.5

1. Evaluate

(a) $\dfrac{3}{4} \div \dfrac{1}{8}$

(b) $\dfrac{8}{9} \div \dfrac{4}{3}$

(c) $\dfrac{-2}{7} \div \dfrac{4}{21}$

(d) $\dfrac{9}{4} \div 1\dfrac{1}{2}$

(e) $\dfrac{5}{6} \div \dfrac{5}{12}$

(f) $\dfrac{99}{100} \div 1\dfrac{4}{5}$

(g) $3\dfrac{1}{4} \div 1\dfrac{1}{8}$

(h) $\left(2\dfrac{1}{4} \div \dfrac{3}{4} \right) \times 2$

(i) $2\dfrac{1}{4} \div \left(\dfrac{3}{4} \times 2 \right)$

(j) $6\dfrac{1}{4} \div 2\dfrac{1}{2} + 5$

(k) $6\dfrac{1}{4} \div \left(2\dfrac{1}{2} + 5 \right)$

Test and assignment exercises 2

1. Evaluate

(a) $\dfrac{3}{4} + \dfrac{1}{6}$

(b) $\dfrac{2}{3} + \dfrac{3}{5} - \dfrac{1}{6}$

(c) $\dfrac{5}{7} - \dfrac{2}{3}$

(d) $2\dfrac{1}{3} - \dfrac{9}{10}$

(e) $5\dfrac{1}{4} + 3\dfrac{1}{6}$

(f) $\dfrac{9}{8} - \dfrac{7}{6} + 1$

(g) $\dfrac{5}{6} - \dfrac{5}{3} + \dfrac{5}{4}$

(h) $\dfrac{4}{5} + \dfrac{1}{3} - \dfrac{3}{4}$

2. Evaluate

(a) $\dfrac{4}{7} \times \dfrac{21}{32}$

(b) $\dfrac{5}{6} \times \dfrac{8}{15}$

(c) $\dfrac{3}{11} \times \dfrac{20}{21}$

(d) $\dfrac{9}{14} \times \dfrac{8}{18}$

(e) $\dfrac{5}{4} \div \dfrac{10}{13}$

(f) $\dfrac{7}{16} \div \dfrac{21}{32}$

(g) $\dfrac{-24}{25} \div \dfrac{51}{50}$

(h) $\dfrac{45}{81} \div \dfrac{25}{27}$

3. Evaluate the following expressions using the BODMAS rule:

(a) $\dfrac{1}{2} + \dfrac{1}{3} \times 2$

(b) $\dfrac{3}{4} \times \dfrac{2}{3} + \dfrac{1}{4}$

(c) $\dfrac{5}{6} \div \dfrac{2}{3} + \dfrac{3}{4}$

(d) $\left(\dfrac{2}{3} + \dfrac{1}{4} \right) \div 4 + \dfrac{3}{5}$

(e) $\left(\dfrac{4}{3} - \dfrac{2}{5} \times \dfrac{1}{3} \right) \times \dfrac{1}{4} + \dfrac{1}{2}$

(f) $\dfrac{3}{4}$ of $\left(1 + \dfrac{2}{3} \right)$

(g) $\dfrac{2}{3}$ of $\dfrac{1}{2} + 1$

(h) $\dfrac{1}{5} \times \dfrac{2}{3} + \dfrac{2}{5} \div \dfrac{4}{5}$

4. Express in their simplest form:

(a) $\dfrac{21}{84}$

(b) $\dfrac{6}{80}$

(c) $\dfrac{34}{85}$

(d) $\dfrac{22}{143}$

(e) $\dfrac{69}{253}$

Decimal fractions

Objectives: This chapter:

- revises the decimal number system
- shows how to write a number to a given number of significant figures
- shows how to write a number to a given number of decimal places

3.1 Decimal numbers

Consider the whole number 478. We can regard it as the sum

$$400 + 70 + 8$$

In this way we see that, in the number 478, the 8 represents eight ones, or 8 units, the 7 represents seven tens, or 70, and the number 4 represents four hundreds or 400. Thus we have the system of hundreds, tens and units familiar from early years in school. All whole numbers can be thought of in this way.

When we wish to deal with proper fractions and mixed fractions, we extend the hundreds, tens and units system as follows. A **decimal point**, '.', marks the end of the whole number part, and the numbers that follow it, to the right, form the fractional part.

A number immediately to the right of the decimal point, that is in the **first decimal place**, represents tenths, so

$$0.1 = \frac{1}{10}$$

$$0.2 = \frac{2}{10} \quad \text{or} \quad \frac{1}{5}$$

$$0.3 = \frac{3}{10} \quad \text{and so on}$$

Note that when there are no whole numbers involved it is usual to write a zero in front of the decimal point, thus, .2 would be written 0.2.

WORKED EXAMPLE

3.1   Express the following decimal numbers as proper fractions in their simplest form

(a) 0.4   (b) 0.5   (c) 0.6

Solution   The first number after the decimal point represents tenths.

(a) $0.4 = \frac{4}{10}$, which simplifies to $\frac{2}{5}$

(b) $0.5 = \frac{5}{10}$ or simply $\frac{1}{2}$

(c) $0.6 = \frac{6}{10} = \frac{3}{5}$

Frequently we will deal with numbers having a whole number part and a fractional part. Thus

$$5.2 = 5 \text{ units} + 2 \text{ tenths}$$

$$= 5 + \frac{2}{10}$$

$$= 5 + \frac{1}{5}$$

$$= 5\frac{1}{5}$$

Similarly,

$$175.8 = 175\frac{8}{10} = 175\frac{4}{5}$$

Numbers in the second position after the decimal point, or the **second decimal place**, represent hundredths, so

$$0.01 = \frac{1}{100}$$

$$0.02 = \frac{2}{100} \quad \text{or} \quad \frac{1}{50}$$

$$0.03 = \frac{3}{100} \quad \text{and so on}$$

Consider 0.25. We can think of this as

$$0.25 = 0.2 + 0.05$$

$$= \frac{2}{10} + \frac{5}{100}$$

$$= \frac{25}{100}$$

We see that 0.25 is equivalent to $\frac{25}{100}$, which in its simplest form is $\frac{1}{4}$.

In fact we can regard any numbers occupying the first two decimal places as hundredths, so that

$$0.25 = \frac{25}{100} \quad \text{or simply } \frac{1}{4}$$

$$0.50 = \frac{50}{100} \quad \text{or } \frac{1}{2}$$

$$0.75 = \frac{75}{100} = \frac{3}{4}$$

WORKED EXAMPLES

3.2   Express the following decimal numbers as proper fractions in their simplest form:

(a) 0.35   (b) 0.56   (c) 0.68

Solution   The first two decimal places represent hundredths:

(a) $0.35 = \frac{35}{100} = \frac{7}{20}$

(b) $0.56 = \frac{56}{100} = \frac{14}{25}$

(c) $0.68 = \frac{68}{100} = \frac{17}{25}$

3.3   Express 37.25 as a mixed fraction in its simplest form.

Solution $37.25 = 37 + 0.25$

$$= 37 + \frac{25}{100}$$

$$= 37 + \frac{1}{4}$$

$$= 37\frac{1}{4}$$

Numbers in the third position after the decimal point, or **third decimal place**, represent thousandths, so

$$0.001 = \frac{1}{1000}$$

$$0.002 = \frac{2}{1000} \quad \text{or} \quad \frac{1}{500}$$

$$0.003 = \frac{3}{1000} \quad \text{and so on}$$

In fact we can regard any numbers occupying the first three positions after the decimal point as thousandths, so that

$$0.356 = \frac{356}{1000} \quad \text{or} \quad \frac{89}{250}$$

$$0.015 = \frac{15}{1000} \quad \text{or} \quad \frac{3}{200}$$

$$0.075 = \frac{75}{1000} = \frac{3}{40}$$

WORKED EXAMPLE

3.4 Write each of the following as a decimal number:

(a) $\frac{3}{10} + \frac{7}{100}$ (b) $\frac{8}{10} + \frac{3}{1000}$

Solution

(a) $\frac{3}{10} + \frac{7}{100} = 0.3 + 0.07 = 0.37$

(b) $\frac{8}{10} + \frac{3}{1000} = 0.8 + 0.003 = 0.803$

You will normally use a calculator to add, subtract, multiply and divide decimal numbers. Generally the more decimal places used, the more accurately we can state a number. This idea is developed in the next section.

Self-assessment questions 3.1

1. State which is the largest and which is the smallest of the following numbers:
23.001, 23.0, 23.00001, 23.0008, 23.01

2. Which is the largest of the following numbers?
0.1, 0.02, 0.003, 0.0004, 0.00005

Exercise 3.1

1. Express the following decimal numbers as proper fractions in their simplest form:
   (a) 0.7   (b) 0.8   (c) 0.9

2. Express the following decimal numbers as proper fractions in their simplest form:
   (a) 0.55   (b) 0.158   (c) 0.98
   (d) 0.099

3. Express each of the following as a mixed fraction in its simplest form:
   (a) 4.6   (b) 5.2   (c) 8.05   (d) 11.59
   (e) 121.09

4. Write each of the following as a decimal number:
   (a) $\frac{6}{10} + \frac{9}{100} + \frac{7}{1000}$ (b) $\frac{8}{100} + \frac{3}{1000}$
   (c) $\frac{17}{1000} + \frac{5}{10}$

3.2 Significant figures and decimal places

The accuracy with which we state a number often depends upon the context in which the number is being used. The volume of a petrol tank is usually given to the nearest litre. It is of no practical use to give such a volume to the nearest cubic centimetre.

When writing a number we often give the accuracy by stating the **number of significant figures** or the **number of decimal places** used. These terms are now explained.

Significant figures

Suppose we are asked to write down the number nearest to 857 using at most two non-zero digits, or numbers. We would write 860. This number is nearer to 857 than any other number with two non-zero digits. We say that 857 to 2 **significant figures** is 860. The words 'significant figures' are usually abbreviated to s.f. Because 860 is larger than 857 we say that the 857 has been **rounded up** to 860.

To write a number to three significant figures we can use no more than three non-zero digits. For example, the number closest to 1784 which has no more than three non-zero digits is 1780. We say that 1784 to 3 significant figures is 1780. In this case, because 1780 is less than 1784 we say that 1784 has been **rounded down** to 1780.

WORKED EXAMPLES

3.5   Write down the number nearest to 86 using only one non-zero digit. Has 86 been rounded up or down?

Solution   The number 86 written to one significant figure is 90. This number is nearer to 86 than any other number having only one non-zero digit. 86 has been rounded up to 90.

3.6   Write down the number nearest to 999 which uses only one non-zero digit.

Solution   The number 999 to one significant figure is 1000. This number is nearer to 999 than any other number having only one non-zero digit.

We now explain the process of writing to a given number of significant figures.

When asked to write a number to, say, three significant figures, 3 s.f., the first step is to look at the first four digits. If asked to write a number to two significant figures we look at the first three digits and so on. We always look at one more digit than the number of significant figures required.

For example, to write 6543.19 to 2 s.f. we would consider the number 6540.00; the digits 3, 1 and 9 are effectively ignored. The next step is to round up or down. If the final digit is a 5 or more then we round up by increasing the previous digit by 1. If the final digit is 4 or less we round down by leaving the previous digit unchanged. Hence when considering 6543.19 to 2 s.f., the 4 in the third place means that we round down to 6500.

To write 23865 to 3 s.f. we would consider the number 23860. The next step is to increase the 8 to a 9. Thus 23865 is rounded up to 23900.

Zeros at the beginning of a number are ignored. To write 0.004693 to 2 s.f. we would first consider the number 0.00469. Note that the zeros at the beginning of the number have not been counted. We then round the 6 to a 7, producing 0.0047.

The following examples illustrate the process.

WORKED EXAMPLES

3.7   Write 36.482 to 3 s.f.

Solution   We consider the first four digits, that is 36.48. The final digit is 8 and so we round up 36.48 to 36.5. To 3 s.f. 36.482 is 36.5.

3.8   Write 1.0049 to 4 s.f.

Solution   To write to 4 s.f. we consider the first five digits, that is 1.0049. The final digit is a 9 and so 1.0049 is rounded up to 1.005.

3.9   Write 695.3 to 2 s.f.

Solution   We consider 695. The final digit is a 5 and so we round up. We cannot round up the 9 to a 10 and so the 69 is rounded up to 70. Hence to 2 s.f. the number is 700.

3.10   Write 0.0473 to 1 s.f.

Solution   We do not count the initial zeros and consider 0.047. The final digit tells us to round up. Hence to 1 s.f. we have 0.05.

3.11   A number is given to 2 s.f. as 67.

(a) What is the maximum value the number could have?
(b) What is the minimum value the number could have?

Solution   (a) To 2 s.f. 67.5 is 68. Any number just below 67.5, for example 67.49 or 67.499, to 2 s.f. is 67. Hence the maximum value of the number is 67.4999....

(b) To 2 s.f. 66.4999... is 66. However, 66.5 to 2 s.f. is 67. The minimum value of the number is thus 66.5.

Decimal places

When asked to write a number to 3 decimal places (3 d.p.) we consider the first 4 decimal places, that is numbers after the decimal point. If asked to write to 2 d.p. we consider the first 3 decimal places and so on. If the final digit is 5 or more we round up, otherwise we round down.

WORKED EXAMPLES

3.12   Write 63.4261 to 2 d.p.

Solution   We consider the number to 3 d.p., that is 63.426. The final digit is 6 and so we round up 63.426 to 63.43. Hence 63.4261 to 2 d.p. is 63.43.

3.13   Write 1.97 to 1 d.p.

Solution   In order to write to 1 d.p. we consider the number to 2 d.p., that is we consider 1.97. The final digit is a 7 and so we round up. The 9 cannot be rounded up and so we look at 1.9. This can be rounded up to 2.0. Hence 1.97 to 1 d.p. is 2.0. Note that it is crucial to write 2.0 and not simply 2, as this shows that the number is written to 1 d.p.

3.14   Write −6.0439 to 2 d.p.

Solution   We consider −6.043. As the final digit is a 3 the number is rounded down to −6.04.

Self-assessment questions 3.2

1. Explain the meaning of 'significant figures'.
2. Explain the process of writing a number to so many decimal places.

Exercise 3.2   *MyMathLab*

1. Write to 3 s.f.
   (a) 6962   (b) 70.406   (c) 0.0123
   (d) 0.010991   (e) 45.607   (f) 2345

2. Write 65.999 to
   (a) 4 s.f.   (b) 3 s.f.   (c) 2 s.f.
   (d) 1 s.f.   (e) 2 d.p.   (f) 1 d.p.

3. Write 9.99 to
   (a) 1 s.f.   (b) 1 d.p.

4. Write 65.4555 to
   (a) 3 d.p.   (b) 2 d.p.   (c) 1 d.p.
   (d) 5 s.f.   (e) 4 s.f.   (f) 3 s.f.   (g) 2 s.f.
   (h) 1 s.f.

Test and assignment exercises 3

1. Express the following numbers as proper fractions in their simplest form:
   (a) 0.74   (b) 0.96   (c) 0.05   (d) 0.25

2. Express each of the following as a mixed fraction in its simplest form:
   (a) 2.5   (b) 3.25   (c) 3.125   (d) 6.875

3. Write each of the following as a decimal number:
   (a) $\frac{3}{10} + \frac{1}{100} + \frac{7}{1000}$ (b) $\frac{5}{1000} + \frac{9}{100}$ (c) $\frac{4}{1000} + \frac{9}{10}$

4. Write 0.09846 to (a) 1 d.p, (b) 2 s.f., (c) 1 s.f.

5. Write 9.513 to (a) 3 s.f., (b) 2 s.f., (c) 1 s.f.

6. Write 19.96 to (a) 1 d.p., (b) 2 s.f., (c) 1 s.f.

Percentage and ratio

4

Objectives: This chapter:

- explains the terms 'percentage' and 'ratio'
- shows how to perform calculations using percentages and ratios
- explains how to calculate the percentage change in a quantity

4.1 Percentage

In everyday life we come across percentages regularly. During sales periods shops offer discounts – for example, we might hear expressions like 'everything reduced by 50%'. Students often receive examination marks in the form of percentages – for example, to achieve a pass grade in a university examination, a student may be required to score at least 40%. Banks and building societies charge interest on loans, and the interest rate quoted is usually given as a percentage, for example 4.75%. Percentages also provide a way of comparing two or more quantities. For example, suppose we want to know which is the better mark: 40 out of 70, or 125 out of 200? By expressing these marks as percentages we will be able to answer this question.

Consequently an understanding of what a percentage is, and an ability to perform calculations involving percentages, are not only useful in mathematical applications, but also essential life skills.

Most calculators have a percentage button and we will illustrate the use of this later in the chapter. However, be aware that different calculators work in different and often confusing ways. Misleading results can be obtained if you do not know how to use your calculator correctly. So it is better if you are not over-reliant on your calculator and instead understand the principles behind percentage calculations.

Fundamentally, a **percentage** is a fraction whose denominator is 100. In fact you can think of the phrase 'per cent' meaning 'out of 100'. We use the symbol % to represent a percentage, as earlier. The following three fractions all have a denominator of 100, and are expressed as percentages as shown:

$$\frac{17}{100} \quad \text{may be expressed as} \quad 17\%$$

$$\frac{50}{100} \quad \text{may be expressed as} \quad 50\%$$

$$\frac{3}{100} \quad \text{may be expressed as} \quad 3\%$$

WORKED EXAMPLE

4.1   Express $\frac{19}{100}$, $\frac{35}{100}$ and $\frac{17.5}{100}$ as percentages.

Solution   All of these fractions have a denominator of 100. So it is straightforward to write down their percentage form:

$$\frac{19}{100} = 19\% \qquad \frac{35}{100} = 35\% \qquad \frac{17.5}{100} = 17.5\%$$

Sometimes it is necessary to convert a fraction whose denominator is not 100, for example $\frac{2}{5}$, into a percentage. This could be done by expressing the fraction as an equivalent fraction with denominator 100, as was explained in Section 2.2 on page 15. However, with calculators readily available, the calculation can be done as follows.

We can use the calculator to divide the numerator of the fraction by the denominator. The answer is then multiplied by 100. The resulting number is the required percentage. So, to convert $\frac{2}{5}$ we perform the following key strokes:

$$2 \div 5 \times 100 = 40$$

and so $\frac{2}{5} = 40\%$. You should check this now using your own calculator,

Key point   To convert a fraction to a percentage, divide the numerator by the denominator, multiply by 100 and then label the result as a percentage.

WORKED EXAMPLES

4.2   Convert $\frac{5}{8}$ into a percentage.

Solution   Using the method described above we find

$$5 \div 8 \times 100 = 62.5$$

Labelling the answer as a percentage, we see that $\frac{5}{8}$ is equivalent to 62.5%.

4.3   Bill scores $\frac{13}{17}$ in a test. In a different test, Mary scores $\frac{14}{19}$. Express the scores as percentages, and thereby make a comparison of the two marks.

Solution   Use your calculator to perform the division and then multiply the result by 100.

Bill's score: $13 \div 17 \times 100 = 76.5$ (1 d.p.)

Mary's score: $14 \div 19 \times 100 = 73.7$ (1 d.p.)

So we see that Bill scores 76.5% and Mary scores 73.7%. Notice that in these percentage forms it is easy to compare the two marks. We see that Bill has achieved the higher score. Making easy comparisons like this is one of the reasons why percentages are used so frequently.

We have seen that percentages are fractions with a denominator of 100, so that, for example, $\frac{19}{100} = 19\%$. Sometimes a fraction may be given not as a numerator divided by a denominator, but in its decimal form. For example, the decimal form of $\frac{19}{100}$ is 0.19. To convert a decimal fraction into a percentage we simply multiply by 100. So

$0.19 = 0.19 \times 100\% = 19\%$

Key point   To convert a decimal fraction to a percentage, multiply by 100 and then label the result as a percentage.

We may also want to reverse the process. Frequently in business calculations involving formulae for interest it is necessary to express a percentage in its decimal form. To convert a percentage to its equivalent decimal form we divide the percentage by 100. Alternatively, using a calculator, input the percentage and press the % button, to convert the percentage to its decimal form.

WORKED EXAMPLE

4.4   Express 50% as a decimal.

Solution   We divide the percentage by 100:

$50 \div 100 = 0.5$

So 50% is equivalent to 0.5. To see why this is the case, remember that 'per cent' literally means 'out of 100' so 50% means 50 out of 100, or $\frac{50}{100}$, or in its simplest form 0.5.

Alternatively, using a calculator, the key strokes

50 %

should give 0.5. Check whether you can do this on your calculator.

| Key point | To convert a percentage to its equivalent decimal fraction form, divide by 100. |

WORKED EXAMPLE

4.5 Express 17.5% as a decimal.

Solution We divide the percentage by 100:

$$17.5 \div 100 = 0.175$$

So 17.5% is equivalent to 0.175. Now check you can obtain the same result using the percentage button on your calculator.

Some percentages appear so frequently in everyday life that it is useful to learn their fraction and decimal fraction equivalent forms.

| Key point | $10\% = 0.1 = \frac{1}{10}$ $25\% = 0.25 = \frac{1}{4}$ |
| | $50\% = 0.5 = \frac{1}{2}$ $75\% = 0.75 = \frac{3}{4}$ $100\% = 1$ |

Recall from Section 1.2 that 'of' means multiply.

We are often asked to calculate a percentage of a quantity: for example, find 17.5% of 160 or 10% of 95. Such calculations arise when finding discounts on prices. Since $17.5\% = \frac{17.5}{100}$ we find

$$17.5\% \text{ of } 160 = \frac{17.5}{100} \times 160 = 28$$

and since $10\% = 0.1$ we may write

$$10\% \text{ of } 95 = 0.1 \times 95 = 9.5$$

Alternatively, the percentage button on a calculator can be used: check you can use your calculator correctly by verifying

17.5 % × 160 = 28

Because finding 10% of a quantity is equivalent to dividing by 10, it is easy to find 10% by moving the decimal point one place to the left.

10 % × 95 = 9.5

4.6   Calculate 27% of 90.

Solution   Using a calculator

$$27 \boxed{\%} \times 90 = 24.3$$

4.7   Calculate 100% of 6.

Solution
$$100 \boxed{\%} \times 6 = 6$$

Observe that 100% of a number is simply the number itself.

4.8   A deposit of £750 increases by 9%. Calculate the resulting deposit.

Solution   We use a calculator to find 9% of 750. This is the amount by which the deposit has increased. Then

$$9 \boxed{\%} \times 750 = 67.50$$

The deposit has increased by £67.50. The resulting deposit is therefore $750 + 67.5 = £817.50$.

Alternatively we may perform the calculation as follows. The original deposit represents 100%. The deposit increases by 9% to 109% of the original. So the resulting deposit is 109% of £750:

$$109 \boxed{\%} \times 750 = £817.50$$

4.9   A television set is advertised at £315. The retailer offers a 10% discount. How much do you pay for the television?

Solution   10% of 315 = 31.50

The discount is £31.50 and so the cost is $315 - 31.5 = £283.50$.

Alternatively we can note that since the discount is 10%, then the selling price is 90% of the advertised price:

$$90 \boxed{\%} \times 315 = 283.50$$

Performing the calculation in the two ways will increase your understanding of percentages and serve as a check.

When a quantity changes, it is sometimes useful to calculate the **percentage change**. For example, suppose a worker earns £14,500 in the current year, and last year earned £13,650. The actual amount earned has changed by $14{,}500 - 13{,}650 = £850$. The percentage change is calculated from the formula:

Key point

$$\text{percentage change} = \frac{\text{change}}{\text{original value}} \times 100 = \frac{\text{new value} - \text{original value}}{\text{original value}} \times 100$$

If the change is positive, then there has been an increase in the measured quantity. If the change is negative, then there has been a decrease in the quantity.

WORKED EXAMPLES

4.10   A worker's earnings increase from £13,650 to £14,500. Calculate the percentage change.

Solution

$$\text{percentage change} = \frac{\text{new value} - \text{original value}}{\text{original value}} \times 100$$

$$= \frac{14{,}500 - 13{,}650}{13{,}650} \times 100$$

$$= 6.23$$

The worker's earnings increased by 6.23%.

4.11   A microwave oven is reduced in price from £149.95 to £135. Calculate the percentage change in price.

Solution

$$\text{percentage change} = \frac{\text{new value} - \text{original value}}{\text{original value}} \times 100$$

$$= \frac{135 - 149.95}{149.95} \times 100$$

$$= -9.97$$

The negative result is indicative of the price decrease. The percentage change in price is approximately –10%.

Self-assessment question 4.1

1. Give one reason why it is sometimes useful to express fractions as percentages.

Exercise 4.1

1. Calculate 23% of 124.

2. Express the following as percentages:

    (a) $\dfrac{9}{11}$ (b) $\dfrac{15}{20}$ (c) $\dfrac{9}{10}$ (d) $\dfrac{45}{50}$ (e) $\dfrac{75}{90}$

3. Express $\frac{13}{12}$ as a percentage.

4. Calculate 217% of 500.

5. A worker earns £400 a week. She receives a 6% increase. Calculate her new weekly wage.

6. A debt of £1200 is decreased by 17%. Calculate the remaining debt.

7. Express the following percentages as decimals:

    (a) 50%   (b) 36%   (c) 75%
    (d) 100%   (e) 12.5%

8. A compact disc player normally priced at £256 is reduced in a sale by 20%. Calculate the sale price.

9. A bank deposit earns 7.5% interest in one year. Calculate the interest earned on a deposit of £15,000.

10. The cost of a car is increased from £6950 to £7495. Calculate the percentage change in price.

11. During a sale, a washing machine is reduced in price from £525 to £399. Calculate the percentage change in price.

4.2 Ratio

Ratios are simply an alternative way of expressing fractions. Consider the problem of dividing £200 between two people, Ann and Bill, in the ratio of $7 : 3$. This means that Ann receives £7 for every £3 that Bill receives. So every £10 is divided as £7 to Ann and £3 to Bill. Thus Ann receives $\frac{7}{10}$ of the money. Now $\frac{7}{10}$ of £200 is $\frac{7}{10} \times 200 = 140$. So Ann receives £140 and Bill receives £60.

WORKED EXAMPLE

4.12   Divide 170 in the ratio $3 : 2$.

Solution A ratio of $3:2$ means that every 5 parts are split as 3 and 2. That is, the first number is $\frac{3}{5}$ of the total; the second number is $\frac{2}{5}$ of the total. So

$$\frac{3}{5} \text{ of } 170 = \frac{3}{5} \times 170 = 102$$

$$\frac{2}{5} \text{ of } 170 = \frac{2}{5} \times 170 = 68$$

The number is divided into 102 and 68.

Note from Worked Example 4.12 that to split a number in a given ratio we first find the total number of parts. The total number of parts is found by adding the numbers in the ratio. For example, if the ratio is given as $m:n$, the total number of parts is $m+n$. Then these $m+n$ parts are split into two with the first number being $\frac{m}{m+n}$ of the total, and the second number being $\frac{n}{m+n}$ of the total. Compare this with Worked Example 4.12.

WORKED EXAMPLE

4.13 Divide 250 cm in the ratio $1:3:4$.

Solution Every 8 cm is divided into 1 cm, 3 cm and 4 cm. Thus the first length is $\frac{1}{8}$ of the total, the second length is $\frac{3}{8}$ of the total, and the third length is $\frac{4}{8}$ of the total:

$$\frac{1}{8} \text{ of } 250 = \frac{1}{8} \times 250 = 31.25$$

$$\frac{3}{8} \text{ of } 250 = \frac{3}{8} \times 250 = 93.75$$

$$\frac{4}{8} \text{ of } 250 = \frac{4}{8} \times 250 = 125$$

The 250 cm length is divided into 31.25 cm, 93.75 cm and 125 cm.

Ratios can be written in different ways. The ratio $3:2$ can also be written as $6:4$. This is clear if we note that $6:4$ is a total of 10 parts split as $\frac{6}{10}$ and $\frac{4}{10}$ of the total. Since $\frac{6}{10}$ is equivalent to $\frac{3}{5}$, and $\frac{4}{10}$ is equivalent to $\frac{2}{5}$, we see that $6:4$ is equivalent to $3:2$.

Generally, any ratio can be expressed as an equivalent ratio by multiplying or dividing each term in the ratio by the same number. So, for example,

$$5 : 3 \text{ is equivalent to } 15 : 9$$

and

$$\frac{3}{4} : 2 \text{ is equivalent to } 3 : 8$$

WORKED EXAMPLES

4.14   Divide a mass of 380 kg in the ratio $\frac{3}{4} : \frac{1}{5}$.

Solution   It is simpler to work with whole numbers, so first of all we produce an equivalent ratio by multiplying each term, first by 4, and then by 5, to give

$$\frac{3}{4} : \frac{1}{5} = 3 : \frac{4}{5} = 15 : 4$$

Note that this is equivalent to multiplying through by the lowest common multiple of 4 and 5.

So dividing 380 kg in the ratio $\frac{3}{4} : \frac{1}{5}$ is equivalent to dividing it in the ratio 15 : 4.

Now the total number of parts is 19 and so we split the 380 kg mass as

$$\frac{15}{19} \times 380 = 300$$

and

$$\frac{4}{19} \times 380 = 80$$

The total mass is split into 300 kg and 80 kg.

4.15   Bell metal, which is a form of bronze, is used for casting bells. It is an alloy of copper and tin. To manufacture bell metal requires 17 parts of copper to every 3 parts of tin.

(a) Express this requirement as a ratio.

(b) Express the amount of tin required as a percentage of the total.

(c) If the total amount of tin in a particular casting is 150 kg, find the amount of copper.

Solution   (a) Copper and tin are needed in the ratio 17 : 3.

(b) $\frac{3}{20}$ of the alloy is tin. Since $\frac{3}{20} = 15\%$ we find that 15% of the alloy is tin.

(c)   A mass of 150 kg of tin makes up 15% of the total. So 1% of the total would have a mass of 10 kg. Copper, which makes up 85%, will have a mass of 850 kg.

Self-assessment question 4.2

1. Dividing a number in the ratio $2:3$ is the same as dividing it in the ratio $10:15$. True or false?

Exercise 4.2

1. Divide 180 in the ratio $8:1:3$.

2. Divide 930 cm in the ratio $1:1:3$.

3. A 6 m length of wood is cut in the ratio $2:3:4$. Calculate the length of each piece.

4. Divide 1200 in the ratio $1:2:3:4$.

5. A sum of £2600 is divided between Alan, Bill and Claire in the ratio of $2\frac{3}{4}:1\frac{1}{2}:2\frac{1}{4}$. Calculate the amount that each receives.

6. A mass of 40 kg is divided into three portions in the ratio $3:4:8$. Calculate the mass of each portion.

7. Express the following ratios in their simplest forms:
    (a) $12:24$ (b) $3:6$ (c) $3:6:12$
    (d) $\frac{1}{3}:7$

8. A box contains two sizes of nails. The ratio of long nails to short nails is $2:7$. Calculate the number of each type if the total number of nails is 108.

Test and assignment exercises 4

1. Express as decimals
    (a) 8%   (b) 18%   (c) 65%

2. Express as percentages
    (a) $\dfrac{3}{8}$ (b) $\dfrac{79}{100}$ (c) $\dfrac{56}{118}$

3. Calculate 27.3% of 1496.

4. Calculate 125% of 125.

5. Calculate 85% of 0.25.

6. Divide 0.5 in the ratio $2:4:9$.

7. A bill totals £234.5 to which is added tax at 17.5%. Calculate the amount of tax to be paid.

8. An inheritance is divided between three people in the ratio $4:7:2$. If the least amount received is £2300 calculate how much the other two people received.

9. Divide 70 in the ratio of $0.5 : 1.3 : 2.1$.

10. Divide 50% in the ratio $2 : 3$.

11. The temperature of a liquid is reduced from 39 °C to 35 °C. Calculate the percentage change in temperature.

12. A jacket priced at £120 is reduced by 30% in a sale. Calculate the sale price of the jacket.

13. The price of a car is reduced from £7250 to £6450. Calculate the percentage change in price.

14. The population of a small town increases from 17296 to 19437 over a five-year period. Calculate the percentage change in population.

15. A number, X, is increased by 20% to form a new number Y. Y is then decreased by 20% to form a third number Z. Express Z in terms of X.

Algebra

5

Objectives

This chapter:

- explains what is meant by 'algebra'
- introduces important algebraic notations
- explains what is meant by a 'power' or 'index'
- illustrates how to evaluate an expression
- explains what is meant by a 'formula'

5.1 What is algebra?

In order to extend the techniques of arithmetic so that they can be more useful in applications we introduce letters or **symbols** to represent quantities of interest. For example, we may choose the capital letter I to stand for the *interest rate* in a business calculation, or the lower case letter t to stand for the *time* in a scientific calculation, and so on. The choice of which letter to use for which quantity is largely up to the user, although some conventions have been developed. Very often the letters x and y are used to stand for arbitrary quantities. **Algebra** is the body of mathematical knowledge that has been developed to manipulate symbols. Some symbols take fixed and unchanging values, and these are known as **constants**. For example, suppose we let the symbol b stand for the boiling point of water. This is fixed at 100 °C and so b is a constant. Some symbols represent quantities that can vary, and these are called **variables**. For example, the velocity of a car might be represented by the symbol v, and might vary from 0 to 100 kilometres per hour.

Algebraic notation

In algebraic work particular attention must be paid to the type of symbol used, so that, for example, the symbol T is quite different from the symbol t.

Table 5.1
The Greek alphabet

A	α	alpha	I	ι	iota	P	ρ	rho
B	β	beta	K	κ	kappa	Σ	σ	sigma
Γ	γ	gamma	Λ	λ	lambda	T	τ	tau
Δ	δ	delta	M	μ	mu	Y	υ	upsilon
E	ε	epsilon	N	ν	nu	Φ	ϕ	phi
Z	ζ	zeta	Ξ	ξ	xi	X	χ	chi
H	η	eta	O	o	omicron	Ψ	ψ	psi
Θ	θ	theta	Π	π	pi	Ω	ω	omega

Your scientific calculator is pre-programmed with the value of π. Check that you can use it.

Usually the symbols chosen are letters from the English alphabet although we frequently meet Greek letters. You may already be aware that the Greek letter 'pi', which has the symbol π, is used in the formula for the area of a circle, and is equal to the constant $3.14159\ldots$. In many calculations π can be approximated by $\frac{22}{7}$. For reference the full Greek alphabet is given in Table 5.1.

Another important feature is the position of a symbol in relation to other symbols. As we shall see in this chapter, the quantities xy, x^y, y^x and x_y all can mean quite different things. When a symbol is placed to the right and slightly higher than another symbol it is referred to as a **superscript**. So the quantity x^y contains the superscript y. Likewise, if a symbol is placed to the right and slightly lower than another symbol it is called a **subscript**. The quantity x_1 contains the subscript 1.

The arithmetic of symbols

Addition $(+)$ If the letters x and y stand for two numbers, their **sum** is written as $x+y$. Note that $x+y$ is the same as $y+x$ just as $4+7$ is the same as $7+4$.

Subtraction $(-)$ The quantity $x-y$ is called the **difference** of x and y, and means the number y subtracted from the number x. Note that $x-y$ is not the same as $y-x$, in the same way that $5-3$ is different from $3-5$.

Multiplication $(\times)$ Five times the number x is written $5 \times x$, although when multiplying the $\times$ sign is sometimes replaced with '$\cdot$', or is even left out altogether. This means that $5 \times x$, $5 \cdot x$ and $5x$ all mean five times the number x. Similarly $x \times y$ can be written $x \cdot y$ or simply xy. When multiplying, the order of the symbols is not important, so that xy is the same as yx just as 5×4 is the same as 4×5. The quantity xy is also known as the **product** of x and y.

Division $(\div)$ $x \div y$ means the number x divided by the number y. This is also written x/y. Here the order is important and x/y is quite different from y/x. An expression involving one symbol divided by another is known as an **algebraic fraction**. The top line is called the **numerator** and the bottom line is called the **denominator**. The quantity x/y is known as the **quotient** of x and y.

A quantity made up of symbols together with $+$, $-$, $\times$ or $\div$ is called an **algebraic expression**. When evaluating an algebraic expression the BODMAS rule given in Chapter 1 applies. This rule reminds us of the correct order in which to evaluate an expression.

Self-assessment questions 5.1

1. Explain what you understand by the term 'algebra'.

2. If m and n are two numbers, explain what is meant by mn.

3. What is an algebraic fraction? Explain the meaning of the terms 'numerator' and 'denominator'.

4. What is the distinction between a superscript and a subscript?

5. What is the distinction between a variable and a constant?

5.2 Powers or indices

Frequently we shall need to multiply a number by itself several times, for example $3 \times 3 \times 3$, or $a \times a \times a \times a$.

To abbreviate such quantities a new notation is introduced. $a \times a \times a$ is written a^3, pronounced 'a cubed'. The superscript 3 is called a **power** or **index** and the letter a is called the **base**. Similarly $a \times a$ is written a^2, pronounced 'a squared' or 'a raised to the power 2'.

The calculator button x^y is used to find powers of numbers.

Most calculators have a button marked x^y, which can be used to evaluate expressions such as 2^8, 3^{11} and so on. Check to see whether your calculator can do these by verifying that $2^8 = 256$ and $3^{11} = 177147$. Note that the plural of index is **indices**.

As a^2 means $a \times a$, and a^3 means $a \times a \times a$, then we interpret a^1 as simply a. That is, any number raised to the power 1 is itself.

Key point   Any number raised to the power 1 is itself, that is $a^1 = a$.

WORKED EXAMPLES

5.1   In the expression 3^8 identify the index and the base.

Solution   In the expression 3^8, the index is 8 and the base is 3.

5.2   Explain what is meant by y^5.

Solution y^5 means $y \times y \times y \times y \times y$.

5.3   Explain what is meant by $x^2 y^3$.

Solution x^2 means $x \times x$; y^3 means $y \times y \times y$. Therefore $x^2 y^3$ means $x \times x \times y \times y \times y$.

5.4   Evaluate 2^3 and 3^4.

Solution 2^3 means $2 \times 2 \times 2$, that is 8. Similarly 3^4 means $3 \times 3 \times 3 \times 3$, that is 81.

5.5   Explain what is meant by 7^1.

Solution   Any number to the power 1 is itself, that is 7^1 is simply 7.

5.6   Evaluate 10^2 and 10^3.

Solution 10^2 means 10×10 or 100. Similarly 10^3 means $10 \times 10 \times 10$ or 1000.

5.7   Use indices to write the expression $a \times a \times b \times b \times b$ more compactly.

Solution $a \times a$ can be written a^2; $b \times b \times b$ can be written b^3. Therefore $a \times a \times b \times b \times b$ can be written as $a^2 \times b^3$ or simply $a^2 b^3$.

5.8   Write out fully $z^3 y^2$.

Solution $z^3 y^2$ means $z \times z \times z \times y \times y$. Note that we could also write this as $zzzyy$.

We now consider how to deal with expressions involving not only powers but other operations as well. Recall from §5.1 that the BODMAS rule tells us the order in which operations should be carried out, but the rule makes no reference to powers. In fact, powers should be given higher priority than any other operation and evaluated first. Consider the expression -4^2. Because the power must be evaluated first -4^2 is equal to -16. On the other hand $(-4)^2$ means $(-4) \times (-4)$ which is equal to $+16$.

WORKED EXAMPLES

5.9   Simplify (a) -5^2,   (b) $(-5)^2$.

Solution   (a) The power is evaluated first. Noting that $5^2 = 25$, we see that $-5^2 = -25$.

Recall that when a negative number is multiplied by another negative number the result is positive.

(b) $(-5)^2$ means $(-5) \times (-5) = +25$.

Note how the brackets can significantly change the meaning of an expression.

5.10   Explain the meanings of $-x^2$ and $(-x)^2$. Are these different?

Solution   In the expression $-x^2$ it is the quantity x that is squared, so that $-x^2 = -(x \times x)$. On the other hand $(-x)^2$ means $(-x) \times (-x)$, which equals $+x^2$. The two expressions are not the same.

Following the previous two examples we emphasise again the importance of the position of brackets in an expression.

Self-assessment questions 5.2

1. Explain the meaning of the terms 'power' and 'base'.

2. What is meant by an index?

3. Explain the distinction between $(xyz)^2$ and xyz^2.

4. Explain the distinction between $(-3)^4$ and -3^4.

Exercise 5.2

1. Evaluate the following without using a calculator: 2^4, $(\frac{1}{2})^2$, 1^8, 3^5 and 0^3.

2. Evaluate 10^4, 10^5 and 10^6 without using a calculator.

3. Use a calculator to evaluate 11^4, 16^8, 39^4 and 1.5^7.

4. Write out fully (a) a^4b^2c and (b) xy^2z^4.

5. Write the following expressions compactly using indices:
   (a) $xxxyyx$ (b) $xxyyzzz$
   (c) $xyzxyz$ (d) $abccba$

6. Using a calculator, evaluate
   (a) 7^4 (b) 7^5 (c) $7^4 \times 7^5$ (d) 7^9
   (e) 8^3 (f) 8^7 (g) $8^3 \times 8^7$ (h) 8^{10}

Can you spot a rule for multiplying numbers with powers?

7. Without using a calculator, find $(-3)^3$, $(-2)^2$, $(-1)^7$ and $(-1)^4$.

8. Use a calculator to find $(-16.5)^3$, $(-18)^2$ and $(-0.5)^5$.

9. Without using a calculator find
   (a) $(-6)^2$ (b) $(-3)^2$ (c) $(-4)^3$
   (d) $(-2)^3$
   Carefully compare your answers with the results of finding -6^2, -3^2, -4^3 and -2^3.

5.3 Substitution and formulae

Substitution means replacing letters by actual numerical values.

WORKED EXAMPLES

5.11 Find the value of a^4 when $a = 3$.

Solution a^4 means $a \times a \times a \times a$. When we **substitute** the number 3 in place of the letter a we find 3^4 or $3 \times 3 \times 3 \times 3$, that is 81.

5.12 Find the value of $a + 7b + 3c$ when $a = 1$, $b = 2$ and $c = 3$.

Solution Letting $b = 2$ we note that $7b = 14$. Letting $c = 3$ we note that $3c = 9$. Therefore, with $a = 1$,

$$a + 7b + 3c = 1 + 14 + 9 = 24$$

5.13 If $x = 4$, find the value of (a) $8x^3$ and (b) $(8x)^3$.

Solution (a) Substituting $x = 4$ into $8x^3$ we find $8 \times 4^3 = 8 \times 64 = 512$.

(b) Substituting $x = 4$ into $(8x)^3$ we obtain $(32)^3 = 32768$. Note that the use of brackets makes a significant difference to the result.

5.14 Evaluate mk, mn and nk when $m = 5$, $n = -4$ and $k = 3$.

Solution $mk = 5 \times 3 = 15$. Similarly $mn = 5 \times (-4) = -20$ and $nk = (-4) \times 3 = -12$.

5.15 Find the value of $-7x$ when (a) $x = 2$ and (b) $x = -2$.

Solution (a) Substituting $x = 2$ into $-7x$ we find -7×2, which equals -14.

(b) Substituting $x = -2$ into $-7x$ we find -7×-2, which equals 14.

5.16 Find the value of x^2 when $x = -3$.

Solution Because x^2 means $x \times x$, its value when $x = -3$ is -3×-3, that is $+9$.

5.17 Find the value of $-x^2$ when $x = -3$.

Solution Recall that a power is evaluated first. So $-x^2$ means $-(x \times x)$. When $x = -3$, this evaluates to $-(-3 \times -3) = -9$.

5.18 Find the value of $x^2 + 3x$ when (a) $x = 2$, (b) $x = -2$.

Solution (a) Letting $x = 2$ we find

$$x^2 + 3x = (2)^2 + 3(2) = 4 + 6 = 10$$

(b) Letting $x = -2$ we find

$$x^2 + 3x = (-2)^2 + 3(-2) = 4 - 6 = -2$$

5.19 Find the value of $\frac{3x^2}{4} + 5x$ when $x = 2$.

Solution   Letting $x = 2$ we find

$$\frac{3x^2}{4} + 5x = \frac{3(2)^2}{4} + 5(2)$$

$$= \frac{12}{4} + 10$$

$$= 13$$

5.20   Find the value of $\frac{x^3}{4}$ when $x = 0.5$.

Solution   When $x = 0.5$ we find

$$\frac{x^3}{4} = \frac{0.5^3}{4} = 0.03125$$

A **formula** is used to relate two or more quantities. You may already be familiar with the common formula used to find the area of a rectangle:

area $=$ length $\times$ breadth

In symbols, writing A for area, l for length and b for breadth we have

$A = l \times b$ or simply $A = lb$

If we are now given particular numerical values for l and b we can use this formula to find A.

WORKED EXAMPLES

5.21   Use the formula $A = lb$ to find A when $l = 10$ and $b = 2.5$.

Solution   Substituting the values $l = 10$ and $b = 2.5$ into the formula $A = lb$ we find $A = 10 \times 2.5 = 25$.

5.22   The formula $V = IR$ is used by electrical engineers. Find the value of V when $I = 12$ and $R = 7$.

Solution   Substituting $I = 12$ and $R = 7$ in $V = IR$ we find $V = 12 \times 7 = 84$.

5.23   Use the formula $y = x^2 + 3x + 4$ to find y when $x = -2$.

Solution   Substituting $x = -2$ into the formula gives

$$y = (-2)^2 + 3(-2) + 4 = 4 - 6 + 4 = 2$$

Self-assessment question 5.3

1. What is the distinction between an algebraic expression and a formula?

Exercise 5.3

1. Evaluate $3x^2y$ when $x = 2$ and $y = 5$.

2. Evaluate $8x + 17y - 2z$ when $x = 6$, $y = 1$ and $z = -2$.

3. The area A of a circle is found from the formula $A = \pi r^2$, where r is the length of the radius. Taking π to be 3.142 find the areas of the circles whose radii, in centimetres, are (a) $r = 10$, (b) $r = 3$, (c) $r = 0.2$.

4. Evaluate $3x^2$ and $(3x)^2$ when $x = 4$.

5. Evaluate $5x^2$ and $(5x)^2$ when $x = -2$.

6. If $y = 4.85$ find
   (a) $7y$ (b) y^2 (c) $5y + 2.5$
   (d) $y^3 - y$

7. If $a = 12.8$, $b = 3.6$ and $c = 9.1$ find
   (a) $a + b + c$ (b) ab (c) bc (d) abc

8. If $C = \frac{5}{9}(F - 32)$, find C when $F = 100$.

9. Evaluate (a) x^2, (b) $-x^2$ and (c) $(-x)^2$, when $x = 7$.

10. Evaluate the following when $x = -2$:
    (a) x^2 (b) $(-x)^2$ (c) $-x^2$
    (d) $3x^2$ (e) $-3x^2$ (f) $(-3x)^2$

11. Evaluate the following when $x = -3$:
    (a) $\frac{x^2}{3}$ (b) $(-x)^2$ (c) $-\left(\frac{x}{3}\right)^2$
    (d) $4x^2$ (e) $-4x^2$ (f) $(-4x)^2$

12. Evaluate $x^2 - 7x + 2$ when $x = -9$.

13. Evaluate $2x^2 + 3x - 11$ when $x = -3$.

14. Evaluate $-x^2 + 3x - 5$ when $x = -1$.

15. Evaluate $-9x^2 + 2x$ when $x = 0$.

16. Evaluate $5x^2 + x + 1$ when (a) $x = 3$, (b) $x = -3$, (c) $x = 0$, (d) $x = -1$.

17. Evaluate $\frac{2x^2}{3} - \frac{x}{2}$ when
    (a) $x = 6$ (b) $x = -6$ (c) $x = 0$
    (d) $x = 1$

18. Evaluate $\frac{4x^2}{5} + 3$ when
    (a) $x = 0$ (b) $x = 1$ (c) $x = 5$
    (d) $x = -5$

19. Evaluate $\frac{x^3}{2}$ when
    (a) $x = -1$ (b) $x = 2$ (c) $x = 4$

20. Use the formula $y = \frac{x^3}{2} + 3x^2$ to find y when
    (a) $x = 0$ (b) $x = 2$ (c) $x = 3$
    (d) $x = -1$

21. If $g = 2t^2 - 1$, find g when
    (a) $t = 3$ (b) $t = 0.5$ (c) $t = -2$

22. In business calculations, the simple interest earned on an investment, I, is calculated from the formula $I = Prn$, where P is the amount invested, r is the interest rate and n is the number of time periods. Evaluate I when
    (a) $P = 15000$, $r = 0.08$ and $n = 5$
    (b) $P = 12500$, $r = 0.075$ and $n = 3$.

23. An investment earning 'compound interest' has a value, S, given by $S = P(1 + r)^n$, where P is the amount invested, r is the interest rate and n is the number of time periods. Calculate S when
    (a) $P = 8250$, $r = 0.05$ and $n = 15$
    (b) $P = 125000$, $r = 0.075$ and $n = 11$.

VIDEO

1. Using a calculator, evaluate 44^3, 0.44^2 and 32.5^3.

2. Write the following compactly using indices:

   (a) $xxxyyyy$ (b) $\dfrac{xxx}{yyyy}$ (c) a^2baab

3. Evaluate the expression $4x^3yz^2$ when $x = 2$, $y = 5$ and $z = 3$.

4. The circumference C of a circle that has a radius of length r is given by the formula $C = 2\pi r$. Find the circumference of the circle with radius 0.5 cm. Take $\pi = 3.142$.

5. Find (a) $21^2 - 16^2$, (b) $(21 - 16)^2$. Comment upon the result.

6. If $x = 4$ and $y = -3$, evaluate

   (a) xy (b) $\dfrac{x}{y}$ (c) $\dfrac{x^2}{y^2}$ (d) $\left(\dfrac{x}{y}\right)^2$

7. Evaluate $2x(x + 4)$ when $x = 7$.

8. Evaluate $4x^2 + 7x$ when $x = 9$.

9. Evaluate $3x^2 - 7x + 12$ when $x = -2$.

10. Evaluate $-x^2 - 11x + 1$ when $x = -3$.

11. The formula $I = V/R$ is used by engineers. Find I when $V = 10$ and $R = 0.01$.

12. Given the formula $A = 1/x$, find A when (a) $x = 1$, (b) $x = 2$, (c) $x = 3$.

13. From the formula $y = 1/(x^2 + x)$ find y when (a) $x = 1$, (b) $x = -1$, (c) $x = 3$.

14. Find the value of $(-1)^n$ (a) when n is an even natural number and (b) when n is an odd natural number. (A natural number is a positive whole number.)

15. Find the value of $(-1)^{n+1}$ (a) when n is an even natural number and (b) when n is an odd natural number.

Indices

6

Objectives: This chapter:

- states three laws used for manipulating indices
- shows how expressions involving indices can be simplified using the three laws
- explains the use of negative powers
- explains square roots, cube roots and fractional powers
- revises multiplication and division by powers of 10
- explains 'scientific notation' for representing very large and very small numbers

6.1 The laws of indices

Recall from Chapter 5 that an index is simply a power and that the plural of index is indices. Expressions involving indices can often be simplified if use is made of the **laws of indices**.

The first law

$$a^m \times a^n = a^{m+n}$$

In words, this states that if two numbers involving the same base but possibly different indices are to be multiplied together, their indices are added. Note that this law can be applied only if both bases are the same.

Key point   The first law: $a^m \times a^n = a^{m+n}$.

WORKED EXAMPLES

6.1   Use the first law of indices to simplify $a^4 \times a^3$.

Solution   Using the first law we have $a^4 \times a^3 = a^{4+3} = a^7$. Note that the same result could be obtained by actually writing out all the terms:

$$a^4 \times a^3 = (a \times a \times a \times a) \times (a \times a \times a) = a^7$$

6.2   Use the first law of indices to simplify $3^4 \times 3^5$.

Solution   From the first law $3^4 \times 3^5 = 3^{4+5} = 3^9$.

6.3   Simplify $a^4 a^7 b^2 b^4$.

Solution $a^4 a^7 b^2 b^4 = a^{4+7} b^{2+4} = a^{11} b^6$. Note that only those quantities with the same base can be combined using the first law.

The second law

$$\frac{a^m}{a^n} = a^{m-n}$$

In words, this states that if two numbers involving the same base but possibly different indices are to be divided, their indices are subtracted.

Key point

The second law: $\dfrac{a^m}{a^n} = a^{m-n}$.

WORKED EXAMPLES

6.4   Use the second law of indices to simplify $\frac{a^5}{a^3}$.

Solution   The second law states that we subtract the indices, that is

$$\frac{a^5}{a^3} = a^{5-3} = a^2$$

6.5   Use the second law of indices to simplify $\frac{3^7}{3^4}$.

Solution   From the second law, $\frac{3^7}{3^4} = 3^{7-4} = 3^3$.

6.6   Using the second law of indices, simplify $\frac{x^3}{x^3}$.

Solution   Using the second law of indices we have $\frac{x^3}{x^3} = x^{3-3} = x^0$. However, note that any expression divided by itself equals 1, and so $\frac{x^3}{x^3}$ must equal 1. We can conclude from this that any number raised to the power 0 equals 1.

Key point   Any number raised to the power 0 equals 1, that is $a^0 = 1$.

WORKED EXAMPLE

6.7   Evaluate (a) 14^0, (b) 0.5^0.

Solution   (a) Any number to the power 0 equals 1 and so $14^0 = 1$.

(b) Similarly, $0.5^0 = 1$.

The third law

$$(a^m)^n = a^{mn}$$

If a number is raised to a power, and the result is itself raised to a power, then the two powers are multiplied together.

Key point   The third law: $(a^m)^n = a^{mn}$.

WORKED EXAMPLES

6.8   Simplify $(3^2)^4$.

Solution   The third law states that the two powers are multiplied:

$$(3^2)^4 = 3^{2 \times 4} = 3^8$$

6.9   Simplify $(x^4)^3$.

Solution   Using the third law:

$$(x^4)^3 = x^{4 \times 3} = x^{12}$$

6.10   Remove the brackets from the expression $(2a^2)^3$.

Solution $(2a^2)^3$ means $(2a^2) \times (2a^2) \times (2a^2)$. We can write this as

$$2 \times 2 \times 2 \times a^2 \times a^2 \times a^2$$

or simply $8a^6$. We could obtain the same result by noting that both terms in the brackets, that is the 2 and the a^2, must be raised to the power 3, that is

$$(2a^2)^3 = 2^3(a^2)^3 = 8a^6$$

The result of the previous example can be generalised to any term of the form $(a^m b^n)^k$. To simplify such an expression we make use of the formula $(a^m b^n)^k = a^{mk} b^{nk}$.

Key point $(a^m b^n)^k = a^{mk} b^{nk}$

WORKED EXAMPLE

6.11   Remove the brackets from the expression $(x^2 y^3)^4$.

Solution   Using the previous result we find

$$(x^2 y^3)^4 = x^8 y^{12}$$

We often need to use several laws of indices in one example.

WORKED EXAMPLES

6.12   Simplify $\frac{(x^3)^4}{x^2}$.

Solution $(x^3)^4 = x^{12}$ using the third law of indices

So

$$\frac{(x^3)^4}{x^2} = \frac{x^{12}}{x^2} = x^{10}$$ using the second law

6.13   Simplify $(t^4)^2 (t^2)^3$.

Solution $(t^4)^2 = t^8$, $(t^2)^3 = t^6$ using the third law

So

$$(t^4)^2 (t^2)^3 = t^8 t^6 = t^{14}$$ using the first law

Self-assessment questions 6.1

1. State the three laws of indices.
2. Explain what is meant by a^0.
3. Explain what is meant by x^1.

Exercise 6.1

1. Simplify
   (a) $5^7 \times 5^{13}$ (b) $9^8 \times 9^5$
   (c) $11^2 \times 11^3 \times 11^4$

2. Simplify
   (a) $\dfrac{15^3}{15^2}$ (b) $\dfrac{4^{18}}{4^9}$ (c) $\dfrac{5^{20}}{5^{19}}$

3. Simplify
   (a) $a^7 a^3$ (b) $a^4 a^5$ (c) $b^{11} b^{10} b$

4. Simplify
   (a) $x^7 \times x^8$ (b) $y^4 \times y^8 \times y^9$

5. Explain why the laws of indices cannot be used to simplify $19^8 \times 17^8$.

6. Simplify
    (a) $(7^3)^2$ (b) $(4^2)^8$ (c) $(7^9)^2$

7. Simplify $\dfrac{1}{(5^3)^8}$.

8. Simplify
    (a) $(x^2 y^3)(x^3 y^2)$ (b) $(a^2 bc^2)(b^2 ca)$

9. Remove the brackets from
    (a) $(x^2 y^4)^5$ (b) $(9x^3)^2$ (c) $(-3x)^3$
    (d) $(-x^2 y^3)^4$

10. Simplify
    (a) $\dfrac{(z^2)^3}{z^3}$ (b) $\dfrac{(y^3)^2}{(y^2)^2}$ (c) $\dfrac{(x^3)^2}{(x^2)^3}$

VIDEO

VIDEO

6.2 Negative powers

Sometimes a number is raised to a negative power. This is interpreted as follows:

$$a^{-m} = \frac{1}{a^m}$$

This can also be rearranged and expressed in the form

$$a^m = \frac{1}{a^{-m}}$$

Key point

$$a^{-m} = \frac{1}{a^m}, \qquad a^m = \frac{1}{a^{-m}}$$

For example,

$$3^{-2} \text{ means } \frac{1}{3^2}, \text{ that is } \frac{1}{9}$$

Similarly,

the number $\dfrac{1}{5^{-2}}$ can be written 5^2, or simply 25

To see the justification for this, note that because any number raised to the power 0 equals 1 we can write

$$\frac{1}{a^m} = \frac{a^0}{a^m}$$

Using the second law of indices to simplify the right-hand side we obtain $\frac{a^0}{a^m} = a^{0-m} = a^{-m}$ so that $\frac{1}{a^m}$ is the same as a^{-m}.

WORKED EXAMPLES

6.14   Evaluate

(a) 2^{-5} (b) $\dfrac{1}{3^{-4}}$

Solution   (a) $2^{-5} = \dfrac{1}{2^5} = \dfrac{1}{32}$ (b) $\dfrac{1}{3^{-4}} = 3^4$ or simply 81

6.15   Evaluate

(a) 10^{-1} (b) 10^{-2}

Solution   (a) 10^{-1} means $\frac{1}{10^1}$, or simply $\frac{1}{10}$. It is important to recognise that 10^{-1} is therefore the same as 0.1.

(b) 10^{-2} means $\frac{1}{10^2}$ or $\frac{1}{100}$. So 10^{-2} is therefore the same as 0.01.

6.16   Rewrite each of the following expressions using only positive powers:

(a) 7^{-3} (b) x^{-5}

Solution   (a) 7^{-3} means the same as $\frac{1}{7^3}$. The expression has now been written using a positive power.

(b) $x^{-5} = \frac{1}{x^5}$.

6.17   Rewrite each of the following expressions using only positive powers:

(a) $\dfrac{1}{x^{-9}}$ (b) $\dfrac{1}{a^{-4}}$

Solution   (a) $\dfrac{1}{x^{-9}} = x^9$ (b) $\dfrac{1}{a^{-4}} = a^4$

6.18   Rewrite each of the following using only negative powers:

(a) 6^8 (b) x^5 (c) z^a

Solution   (a) $6^8 = \dfrac{1}{6^{-8}}$ (b) $x^5 = \dfrac{1}{x^{-5}}$ (c) $z^a = \dfrac{1}{z^{-a}}$

6.19   Simplify

(a) $x^{-2}x^7$ (b) $\dfrac{x^{-3}}{x^{-5}}$

Solution   (a) To simplify $x^{-2}x^7$ we can use the first law of indices to write it as $x^{-2+7} = x^5$.

(b) To simplify $\dfrac{x^{-3}}{x^{-5}}$ we can use the second law of indices to write it as $x^{-3-(-5)} = x^{-3+5} = x^2$.

6.20   Simplify

(a) $(x^{-3})^5$ (b) $\dfrac{1}{(x^{-2})^2}$

Solution (a) To simplify $(x^{-3})^5$ we can use the third law of indices and write it as $x^{-3 \times 5} = x^{-15}$. The answer could also be written as $\frac{1}{x^{15}}$.

(b) Note that $(x^{-2})^2 = x^{-4}$ using the third law. So $\frac{1}{(x^{-2})^2} = \frac{1}{x^{-4}}$. This could also be written as x^4.

Self-assessment question 6.2

1. Explain how the negative power in a^{-m} is interpreted.

Exercise 6.2

1. Without using a calculator express each of the following as a proper fraction:
    (a) 2^{-2} (b) 2^{-3} (c) 3^{-2} (d) 3^{-3}
    (e) 5^{-2} (f) 4^{-2} (g) 9^{-1} (h) 11^{-2}
    (i) 7^{-1}

2. Express each of the following as decimal fractions:
    (a) 10^{-1} (b) 10^{-2} (c) 10^{-6} (d) $\frac{1}{10^2}$
    (e) $\frac{1}{10^3}$ (f) $\frac{1}{10^4}$

3. Write each of the following using only a positive power:
    (a) x^{-4} (b) $\frac{1}{x^{-5}}$ (c) x^{-7} (d) y^{-2}
    (e) $\frac{1}{y^{-1}}$ (f) y^{-1} (g) y^{-2} (h) z^{-1}
    (i) $\frac{1}{z^{-1}}$

4. Simplify the following using the laws of indices and write your results using only positive powers:
    (a) $x^{-2}x^{-1}$ (b) $x^{-3}x^{-2}$ (c) x^3x^{-4}
    (d) $x^{-4}x^9$ (e) $\frac{x^{-2}}{x^{11}}$ (f) $(x^{-4})^2$
    (g) $(x^{-3})^3$ (h) $(x^2)^{-2}$

5. Simplify
    (a) $a^{13}a^{-2}$ (b) $x^{-9}x^{-7}$ (c) $x^{-21}x^2x$
    (d) $(4^{-3})^2$

6. Evaluate
    (a) 10^{-3} (b) 10^{-4} (c) 10^{-5}

7. Evaluate $4^{-8}/4^{-6}$ and $3^{-5}/3^{-8}$ without using a calculator.

6.3 Square roots, cube roots and fractional powers

Square roots

Consider the relationship between the numbers 5 and 25. We know that $5^2 = 25$ and so 25 is the square of 5. Equivalently we say that 5 is a **square root** of 25. The symbol $\sqrt[2]{\ }$, or simply $\sqrt{\ }$, is used to denote a square root and we write

$$5 = \sqrt{25}$$

We can picture this as follows:

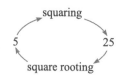

From this we see that taking the square root can be thought of as reversing the process of squaring.

We also note that

$$(-5) \times (-5) = (-5)^2$$
$$= 25$$

and so -5 is also a square root of 25. Hence we can write

$$-5 = \sqrt{25}$$

We can write both results together by using the 'plus or minus' sign $\pm$. We write

$$\sqrt{25} = \pm 5$$

In general, a **square root** of a number is a number that when squared gives the original number. Note that there are two square roots of any positive number but negative numbers possess no square roots.

Most calculators enable you to find square roots although only the positive value is normally given. Look for a $\sqrt{}$ or 'sqrt' button on your calculator.

WORKED EXAMPLE

6.21   (a)   Use your calculator to find $\sqrt{79}$ correct to 4 decimal places.

(b)   Check your answers are correct by squaring them.

Solution   (a)   Using the $\sqrt{}$ button on the calculator you should verify that

$$\sqrt{79} = 8.8882 \text{ (to 4 decimal places)}$$

The second square root is -8.8882. Thus we can write

$$\sqrt{79} = \pm 8.8882$$

(b)   Squaring either of the numbers ± 8.8882 we recover the original number, 79.

Cube roots

The **cube root** of a number is a number that when cubed gives the original number. The symbol for a cube root is $\sqrt[3]{}$. So, for example, since $2^3 = 8$ we can write $\sqrt[3]{8} = 2$.

We can picture this as follows:

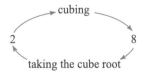

We can think of taking the cube root as reversing the process of cubing. As another example we note that $(-2)^3 = -8$ and hence $\sqrt[3]{-8} = -2$. All numbers, both positive and negative, possess a single cube root.

Your calculator may enable you to find a cube root. Look for a button marked $\sqrt[3]{}$. If so, check that you can use it correctly by verifying that

$$\sqrt[3]{46} = 3.5830$$

Fourth, fifth and other roots are defined in a similar way. For example, since

$$8^5 = 32768$$

we can write

$$\sqrt[5]{32768} = 8$$

Fractional powers

Sometimes fractional powers are used. The following example helps us to interpret a fractional power.

WORKED EXAMPLE

6.22   Simplify

(a) $x^{\frac{1}{2}}x^{\frac{1}{2}}$ (b) $x^{\frac{1}{3}}x^{\frac{1}{3}}x^{\frac{1}{3}}$

Use your results to interpret the fractional powers $\frac{1}{2}$ and $\frac{1}{3}$.

Solution   (a) Using the first law we can write

$$x^{\frac{1}{2}}x^{\frac{1}{2}} = x^{\frac{1}{2}+\frac{1}{2}} = x^1 = x$$

(b) Similarly,

$$x^{\frac{1}{3}}x^{\frac{1}{3}}x^{\frac{1}{3}} = x^{\frac{1}{3}+\frac{1}{3}+\frac{1}{3}} = x^1 = x$$

From (a) we see that

$$(x^{\frac{1}{2}})^2 = x$$

So when $x^{\frac{1}{2}}$ is squared, the result is x. Thus $x^{\frac{1}{2}}$ is simply the square root of x, that is

$$x^{\frac{1}{2}} = \sqrt{x}$$

Similarly, from (b)

$$(x^{\frac{1}{3}})^3 = x$$

and so $x^{\frac{1}{3}}$ is the cube root of x, that is

$$x^{\frac{1}{3}} = \sqrt[3]{x}$$

| Key point | $x^{\frac{1}{2}} = \sqrt{x}, \qquad x^{\frac{1}{3}} = \sqrt[3]{x}$ |

More generally we have the following result:

| Key point | $x^{\frac{1}{n}} = \sqrt[n]{x}$ |

Your scientific calculator will probably be able to find fractional powers. The button may be marked $x^{1/y}$ or $\sqrt[y]{x}$. Check that you can use it correctly by working through the following examples.

WORKED EXAMPLES

6.23 Evaluate to 3 decimal places, using a calculator:

(a) $3^{\frac{1}{4}}$ (b) $15^{1/5}$

Solution Use your calculator to obtain the following solutions:
(a) 1.316   (b) 1.719
Note in part (a) that although the calculator gives just a single fourth root, there is another, -1.316.

6.24 Evaluate $(81)^{1/2}$.

Solution $(81)^{1/2} = \sqrt{81} = \pm 9$.

6.25 Explain what is meant by the number $27^{1/3}$.

Solution $27^{1/3}$ can be written $\sqrt[3]{27}$, that is the cube root of 27. The cube root of 27 is 3, since $3 \times 3 \times 3 = 27$, and so $27^{1/3} = 3$. Note also that since $27 = 3^3$ we can write

$$(27)^{1/3} = (3^3)^{1/3} = 3^{(3 \times 1/3)} \quad \text{using the third law}$$
$$= 3^1 \quad = 3$$

The following worked example shows how we deal with negative fractional powers.

6.26 Explain what is meant by the number $(81)^{-1/2}$.

Solution Recall from our work on negative powers that $a^{-m} = 1/a^m$. Therefore we can write $(81)^{-1/2}$ as $1/(81)^{1/2}$. Now $81^{1/2} = \sqrt{81} = \pm 9$ and so

$$(81)^{-1/2} = \frac{1}{\pm 9} = \pm \frac{1}{9}$$

6.27   Write each of the following using a single index:

(a) $(5^2)^{\frac{1}{3}}$ (b) $(5^{-2})^{\frac{1}{3}}$

Solution   (a) Using the third law of indices we find

$$(5^2)^{\frac{1}{3}} = 5^{2 \times \frac{1}{3}} = 5^{\frac{2}{3}}$$

Note that $(5^2)^{\frac{1}{3}}$ is the cube root of 5^2, that is $\sqrt[3]{25}$ or 2.9240.

(b) Using the third law of indices we find

$$(5^{-2})^{\frac{1}{3}} = 5^{-2 \times \frac{1}{3}} = 5^{-\frac{2}{3}}$$

Note that there is a variety of equivalent ways in which this can be expressed, for example $\sqrt[3]{\frac{1}{5^2}}$ or $\sqrt[3]{\frac{1}{25}}$, or as $\frac{1}{5^{2/3}}$.

6.28   Write each of the following using a single index:

(a) $\sqrt{x^3}$ (b) $(\sqrt{x})^3$

Solution   (a) Because the square root of a number can be expressed as that number raised to the power $\frac{1}{2}$ we can write

$$\sqrt{x^3} = (x^3)^{\frac{1}{2}}$$

$$= x^{3 \times \frac{1}{2}} \quad \text{using the third law}$$

$$= x^{\frac{3}{2}}$$

(b) $(\sqrt{x})^3 = (x^{\frac{1}{2}})^3$

$$= x^{\frac{3}{2}} \quad \text{using the third law}$$

Note from this example that $\sqrt{x^3} = (\sqrt{x})^3$.

Note that by generalising the results of the two previous worked examples we have the following:

Key point $a^{\frac{m}{n}} = \sqrt[n]{a^m} = (\sqrt[n]{a})^m$

Self-assessment questions 6.3

1. Explain the meaning of the fractional powers $x^{1/2}$ and $x^{1/3}$.

2. What are the square roots of 100? Explain why the number -100 does not have any square roots.

Exercise 6.3

1. Evaluate
   (a) $64^{1/3}$ (b) $144^{1/2}$ (c) $16^{-1/4}$
   (d) $25^{-1/2}$ (e) $\dfrac{1}{32^{-1/5}}$

2. Simplify and then evaluate
   (a) $(3^{-1/2})^4$ (b) $(8^{1/3})^{-1}$

3. Write each of the following using a single index:
   (a) $\sqrt{8}$ (b) $\sqrt[3]{12}$ (c) $\sqrt[4]{16}$ (d) $\sqrt{13^3}$
   (e) $\sqrt[3]{4^7}$

4. Write each of the following using a single index:
   (a) $\sqrt{x}$ (b) $\sqrt[3]{y}$ (c) $\sqrt[4]{x^5}$ (d) $\sqrt[3]{5^7}$

6.4 Multiplication and division by powers of 10

To multiply and divide decimal fractions by powers of 10 is particularly simple. For example, to multiply 256.875 by 10 the decimal point is moved one place to the right, that is

$$256.875 \times 10 = 2568.75$$

To multiply by 100 the decimal point is moved two places to the right. So

$$256.875 \times 100 = 25687.5$$

To divide a number by 10, the decimal point is moved one place to the left. This is equivalent to multiplying by 10^{-1}. To divide by 100, the decimal point is moved two places to the left. This is equivalent to multiplying by 10^{-2}.

In general, to multiply a number by 10^n, the decimal point is moved n places to the right if n is a positive integer, and n places to the left if n is a negative integer. If necessary, additional zeros are inserted to make up the required number of digits. Consider the following example.

WORKED EXAMPLE

6.29 Without the use of a calculator, write down

(a) 75.45×10^3 (b) 0.056×10^{-2} (c) 96.3×10^{-3} (d) 0.00743×10^5

Solution (a) The decimal point is moved three places to the right: $75.45 \times 10^3 = 75450$. It has been necessary to include an additional zero to make up the required number of digits.

(b) The decimal point is moved two places to the left: $0.056 \times 10^{-2} = 0.00056$.

(c) $96.3 \times 10^{-3} = 0.0963$.

(d) $0.00743 \times 10^5 = 743$.

Exercise 6.4

1. Without the use of a calculator write down:
   (a) 7.43×10^2 (b) 7.43×10^4 (c) 0.007×10^4 (d) 0.07×10^{-2}

2. Write each of the following as a multiple of 10^2:
   (a) 300 (b) 356 (c) 32 (d) 0.57

6.5 Scientific notation

It is often necessary to use very large numbers such as 65000000000 or very small numbers such as 0.000000001. **Scientific notation** can be used to express such numbers in a more concise form, which avoids writing very lengthy strings of numbers. Each number is written in the form

$$a \times 10^n$$

where a is usually a number between 1 and 10. We also make use of the fact that

$$10 = 10^1, \qquad 100 = 10^2, \qquad 1000 = 10^3 \text{ and so on}$$

and also that

$$10^{-1} = \frac{1}{10} = 0.1, \qquad 10^{-2} = \frac{1}{100} = 0.01 \text{ and so on}$$

Then, for example,

the number 4000 can be written $4 \times 1000 = 4 \times 10^3$

Similarly

the number 68000 can be written $6.8 \times 10000 = 6.8 \times 10^4$

and

the number 0.09 can be written $9 \times 0.01 = 9 \times 10^{-2}$

Note that all three numbers have been written in the form $a \times 10^n$ where a lies between 1 and 10.

WORKED EXAMPLES

6.30 Express the following numbers in scientific notation:

(a) 54 (b) -276 (c) 0.3

Solution   (a)   54 can be written as 5.4×10, so in scientific notation we have 5.4×10^1.

(b)   Negative numbers cause no problem: $-276 = -2.76 \times 10^2$.

(c)   We can write 0.3 as 3×0.1 or 3×10^{-1}.

6.31   Write out fully the following numbers:

(a) 2.7×10^{-1} (b) 9.6×10^5 (c) -8.2×10^2

Solution   (a) $2.7 \times 10^{-1} = 0.27$.

(b) $9.6 \times 10^5 = 9.6 \times 100000 = 960000$.

(c) $-8.2 \times 10^2 = -8.2 \times 100 = -820$.

6.32   Simplify the expression $(3 \times 10^2) \times (5 \times 10^3)$.

Solution   The order in which the numbers are written down does not matter, and so we can write

$$(3 \times 10^2) \times (5 \times 10^3) = 3 \times 5 \times 10^2 \times 10^3 = 15 \times 10^5$$

Noting that $15 = 1.5 \times 10$ we can express the final answer in scientific notation:

$$15 \times 10^5 = 1.5 \times 10 \times 10^5 = 1.5 \times 10^6$$

Hence

$$(3 \times 10^2) \times (5 \times 10^3) = 1.5 \times 10^6$$

Self-assessment question 6.5

1. What is the purpose of using scientific notation?

Exercise 6.5

1. Express each of the following numbers in scientific notation:
   (a) 45   (b) 45000   (c) -450 (d) 90000000   (e) 0.15   (f) 0.00036   (g) 3.5
   (h) -13.2 (i) 1000000   (j) 0.0975   (k) 45.34

2. Write out fully the following numbers:
   (a) 3.75×10^2 (b) 3.97×10^1 (c) 1.875×10^{-1} (d) -8.75×10^{-3}

3. Simplify each of the following expressions, writing your final answer in scientific notation:
   (a) $(4 \times 10^3) \times (6 \times 10^4)$ (b) $(9.6 \times 10^4) \times (8.3 \times 10^3)$ (c) $(1.2 \times 10^{-3}) \times (8.7 \times 10^{-2})$
   (d) $\dfrac{9.37 \times 10^4}{6.14 \times 10^5}$ (e) $\dfrac{4.96 \times 10^{-2}}{9.37 \times 10^{-5}}$

Test and assignment exercises 6

1. Simplify
   (a) $\dfrac{z^5}{z^{-5}}$ (b) z^0 (c) $\dfrac{z^8 z^6}{z^{14}}$

2. Evaluate
   (a) $0.25^{1/2}$ (b) $(4096)^{1/3}$ (c) $(2601)^{1/2}$ (d) $16^{-1/2}$

3. Simplify $\dfrac{x^8 x^{-3}}{x^{-5} x^2}$.

4. Find the value of $(1/7)^0$.

5. Remove the brackets from
   (a) $(abc^2)^2$ (b) $(xy^2 z^3)^2$ (c) $(8x^2)^{-3}$

6. Express each of the following numbers in scientific notation:
   (a) 5792   (b) 98.4   (c) 0.001   (d) −66.667

Simplifying algebraic expressions

7

Objectives: This chapter:

- describes a number of ways in which complicated algebraic expressions can be simplified

7.1 Addition and subtraction of like terms

Like terms are multiples of the same quantity. For example, $3y$, $72y$ and $0.5y$ are all multiples of y and so are like terms. Similarly, $5x^2$, $-3x^2$ and $\frac{1}{2}x^2$ are all multiples of x^2 and so are like terms. xy, $17xy$ and $-91xy$ are all multiples of xy and are therefore like terms. Like terms can be collected together and added or subtracted in order to simplify them.

WORKED EXAMPLES

7.1   Simplify $3x + 7x - 2x$.

Solution   All three terms are multiples of x and so are like terms. Therefore $3x + 7x - 2x = 8x$.

7.2   Simplify $3x + 2y$.

Solution $3x$ and $2y$ are not like terms. One is a multiple of x and the other is a multiple of y. The expression $3x + 2y$ cannot be simplified.

7.3   Simplify $x + 7x + x^2$.

Solution   The like terms are x and $7x$. These can be simplified to $8x$. Then $x + 7x + x^2 = 8x + x^2$. Note that $8x$ and x^2 are not like terms and so this expression cannot be simplified further.

7.4   Simplify $ab + a^2 - 7b^2 + 9ab + 8b^2$.

Solution   The terms ab and $9ab$ are like terms. Similarly the terms $-7b^2$ and $8b^2$ are like terms. These can be collected together and then added or subtracted as appropriate. Thus

$$ab + a^2 - 7b^2 + 9ab + 8b^2 = ab + 9ab + a^2 - 7b^2 + 8b^2$$
$$= 10ab + a^2 + b^2$$

Exercise 7.1

1. Simplify, if possible,
   (a) $5p - 10p + 11q + 8q$ (b) $-7r - 13s + 2r + z$ (c) $181z + 13r - 2$
   (d) $x^2 + 3y^2 - 2y + 7x^2$ (e) $4x^2 - 3x + 2x + 9$

2. Simplify
   (a) $5y + 8p - 17y + 9q$ (b) $7x^2 - 11x^3 + 14x^2 + y^3$ (c) $4xy + 3xy + y^2$
   (d) $xy + yx$ (e) $xy - yx$

7.2 Multiplying algebraic expressions and removing brackets

Recall that when multiplying two numbers together the order in which we write them is irrelevant. For example, both 5×4 and 4×5 equal 20.

When multiplying three or more numbers together the order in which we carry out the multiplication is also irrelevant. By this we mean, for example, that when asked to multiply $3 \times 4 \times 5$ we can think of this as either $(3 \times 4) \times 5$ or as $3 \times (4 \times 5)$. Check for yourself that the result is the same, 60, either way.

It is also important to appreciate that $3 \times 4 \times 5$ could have been written as $(3)(4)(5)$.

It is essential that you grasp these simple facts about numbers in order to understand the algebra that follows. This is because identical rules are applied. Rules for determining the sign of the answer when multiplying positive and negative algebraic expressions are also the same as those used for multiplying numbers.

Key point   When multiplying

$$\text{positive} \times \text{positive} = \text{positive}$$
$$\text{positive} \times \text{negative} = \text{negative}$$
$$\text{negative} \times \text{positive} = \text{negative}$$
$$\text{negative} \times \text{negative} = \text{positive}$$

We introduce the processes involved in removing brackets using some simple examples.

WORKED EXAMPLES

7.5   Simplify $3(4x)$.

Solution   Just as with numbers $3(4x)$ could be written as $3 \times (4 \times x)$, and then as $(3 \times 4) \times x$, which evaluates to $12x$.
So $3(4x) = 12x$.

7.6   Simplify $5(3y)$.

Solution $5(3y) = 5 \times 3 \times y = 15y$.

7.7   Simplify $(5a)(3a)$.

Solution   Here we can write $(5a)(3a) = (5 \times a) \times (3 \times a)$. Neither the order in which we carry out the multiplications nor the order in which we write down the terms matters, and so we can write this as

$$(5a)(3a) = (5 \times 3)(a \times a)$$

As we have shown, it is usual to write numbers at the beginning of an expression. This simplifies to $15 \times a^2$, that is $15a^2$. Hence

$$(5a)(3a) = 15a^2$$

7.8   Simplify $4x^2 \times 7x^5$.

Solution   Recall that, when multiplying, the order in which we write down the terms does not matter. Therefore we can write

$$4x^2 \times 7x^5 = 4 \times 7 \times x^2 \times x^5$$

which equals $28x^{2+5} = 28x^7$.

7.9   Simplify $7(2b^2)$.

Solution $7(2b^2) = 7 \times (2 \times b^2) = (7 \times 2) \times b^2 = 14b^2$.

7.10   Simplify $(a) \times (-b)$

Solution   Here we have the product of a positive and a negative quantity. The result will be negative. We write

$$(a) \times (-b) = -ab$$

7.11   Explain the distinction between ab^2 and $(ab)^2$.

Solution ab^2 means $a \times b \times b$ whereas $(ab)^2$ means $(ab) \times (ab)$ which equals $a \times b \times a \times b$. The latter could also be written as a^2b^2.

7.12   Simplify (a) $(6z)(8z)$, (b) $(6z) + (8z)$, noting the distinction between the two results.

Solution   (a) $(6z)(8z) = 48z^2$.

(b) $(6z) + (8z)$ is the addition of like terms. This simplifies to $14z$.

7.13   Simplify (a) $(6x)(-2x)$, (b) $(-3y^2)(-2y)$.

Solution   (a) $(6x)(-2x)$ means $(6x) \times (-2x)$, which equals $-12x^2$.

(b) $(-3y^2)(-2y) = (-3y^2) \times (-2y) = 6y^3$.

Self-assessment questions 7.2

1. Two negative expressions are multiplied together. State the sign of the resulting product.

2. Three negative expressions are multiplied together. State the sign of the resulting product.

Exercise 7.2

1. Simplify each of the following:
   (a) $(4)(3)(7)$ (b) $(7)(4)(3)$ (c) $(3)(4)(7)$

2. Simplify
   (a) $5 \times (4 \times 2)$ (b) $(5 \times 4) \times 2$

3. Simplify each of the following:
   (a) $7(2z)$ (b) $15(2y)$ (c) $(2)(3)x$
   (d) $9(3a)$ (e) $(11)(5a)$ (f) $2(3x)$

4. Simplify each of the following:
   (a) $5(4x^2)$ (b) $3(2y^3)$ (c) $11(2u^2)$
   (d) $(2 \times 4) \times u^2$ (e) $(13)(2z^2)$

5. Simplify
   (a) $(7x)(3x)$ (b) $3a(7a)$ (c) $14a(a)$

6. Simplify
   (a) $5y(3y)$ (b) $5y + 3y$
   Explain why the two results are not the same.

7. Simplify the following:
   (a) $(abc)(a^2bc)$ (b) $x^2y(xy)$
   (c) $(xy^2)(xy^2)$

8. Explain the distinction, if any, between $(xy^2)(xy^2)$ and xy^2xy^2.

9. Explain the distinction, if any, between $(xy^2)(xy^2)$ and $(xy^2) + (xy^2)$.
   In both cases simplify the expressions.

10. Simplify
    (a) $(3z)(-7z)$ (b) $3z - 7z$

11. Simplify
    (a) $(-x)(3x)$ (b) $-x + 3x$

12. Simplify
    (a) $(-2x)(-x)$ (b) $-2x - x$

7.3 Removing brackets from $a(b + c)$, $a(b - c)$ and $(a + b)(c + d)$

Recall from your study of arithmetic that the expression $(5 - 4) + 7$ is different from $5 - (4 + 7)$ because of the position of the brackets. In order to simplify an expression it is often necessary to remove brackets.

Removing brackets from expressions of the form $a(b + c)$ and $a(b - c)$

In an expression such as $a(b + c)$, it is intended that the a multiplies all the bracketed terms:

Key point	$a(b + c) = ab + ac$ Similarly: $a(b - c) = ab - ac$

WORKED EXAMPLES

7.14   Remove the brackets from

(a) $6(x + 5)$ (b) $8(2x - 4)$

Solution   (a) In the expression $6(x + 5)$ it is intended that the 6 multiplies both terms in the brackets. Therefore

$$6(x + 5) = 6x + 30$$

(b) In the expression $8(2x - 4)$ the 8 multiplies both terms in the brackets so that

$$8(2x - 4) = 16x - 32$$

7.15   Remove the brackets from the expression $7(5x + 3y)$.

Solution   The 7 multiplies both the terms in the bracket. Therefore

$$7(5x + 3y) = 7(5x) + 7(3y) = 35x + 21y$$

7.16   Remove the brackets from $-(x + y)$.

Solution   The expression $-(x + y)$ actually means $-1(x + y)$. It is intended that the -1 multiplies both terms in the brackets, therefore

$$-(x + y) = -1(x + y) = (-1) \times x + (-1) \times y = -x - y$$

7.17   Remove the brackets from the expression

$$(x + y)z$$

Solution   Note that the order in which we write down the terms to be multiplied does not matter, so that we can write $(x + y)z$ as $z(x + y)$. Then

$$z(x + y) = zx + zy$$

Alternatively note that $(x + y)z = xz + yz$, which is an equivalent form of the answer.

7.18   Remove the brackets from the expressions

VIDEO

(a) $5(x - 2y)$ (b) $(x + 3)(-1)$

Solution   (a) $5(x - 2y) = 5x - 5(2y) = 5x - 10y$.

(b) $(x + 3)(-1) = (-1)(x + 3) = -1x - 3 = -x - 3$.

7.19   Simplify $x + 8(x - y)$.

Solution   An expression such as this is simplified by first removing the brackets and then collecting together like terms. Removing the brackets we find

$$x + 8(x - y) = x + 8x - 8y$$

Collecting like terms we obtain $9x - 8y$.

7.20   Remove the brackets from

(a) $\frac{1}{2}(x + 2)$ (b) $\frac{1}{2}(x - 2)$ (c) $-\frac{1}{3}(a + b)$

Solution   (a) In the expression $\frac{1}{2}(x + 2)$ it is intended that the $\frac{1}{2}$ multiplies both the terms in the brackets. So

$$\frac{1}{2}(x + 2) = \frac{1}{2}x + \frac{1}{2}(2) = \frac{1}{2}x + 1$$

(b) Similarly,

$$\frac{1}{2}(x - 2) = \frac{1}{2}x - \frac{1}{2}(2) = \frac{1}{2}x - 1$$

(c) In the expression $-\frac{1}{3}(a + b)$ the term $-\frac{1}{3}$ multiplies both terms in the brackets. So

$$-\frac{1}{3}(a + b) = -\frac{1}{3}a - \frac{1}{3}b$$

Removing brackets from expressions of the form $(a + b)(c + d)$

In the expression $(a + b)(c + d)$ it is intended that the quantity $(a + b)$ multiplies both the c and the d in the second brackets. Therefore

$$(a + b)(c + d) = (a + b)c + (a + b)d$$

Each of these two terms can be expanded further to give

$$(a + b)c = ac + bc \quad \text{and} \quad (a + b)d = ad + bd$$

Therefore

Key point $(a + b)(c + d) = ac + bc + ad + bd$

WORKED EXAMPLES

7.21   Remove the brackets from $(3 + x)(2 + y)$.

Solution $(3 + x)(2 + y) = (3 + x)(2) + (3 + x)y$

$$= 6 + 2x + 3y + xy$$

7.22   Remove the brackets from $(x + 6)(x - 3)$.

Solution $(x + 6)(x - 3) = (x + 6)x + (x + 6)(-3)$

$$= x^2 + 6x - 3x - 18$$

$$= x^2 + 3x - 18$$

7.23   Remove the brackets from

(a) $(1 - x)(2 - x)$ (b) $(-x - 2)(2x - 1)$

Solution   (a) $(1 - x)(2 - x) = (1 - x)2 + (1 - x)(-x)$

$$= 2 - 2x - x + x^2$$

$$= 2 - 3x + x^2$$

(b) $(-x - 2)(2x - 1) = (-x - 2)(2x) + (-x - 2)(-1)$

$$= -2x^2 - 4x + x + 2$$

$$= -2x^2 - 3x + 2$$

7.24   Remove the brackets from the expression $3(x + 1)(x - 1)$.

VIDEO

Solution   First consider the expression $(x + 1)(x - 1)$:

$$(x + 1)(x - 1) = (x + 1)x + (x + 1)(-1)$$

$$= x^2 + x - x - 1$$

$$= x^2 - 1$$

Then $3(x + 1)(x - 1) = 3(x^2 - 1) = 3x^2 - 3$.

Exercise 7.3

1. Remove the brackets from
   (a) $4(x + 1)$ (b) $-4(x + 1)$
   (c) $4(x - 1)$ (d) $-4(x - 1)$

2. Remove the brackets from the following
   expressions:
   (a) $5(x - y)$ (b) $19(x + 3y)$
   (c) $8(a + b)$ (d) $(5 + x)y$
   (e) $12(x + 4)$ (f) $17(x - 9)$
   (g) $-(a - 2b)$ (h) $\frac{1}{2}(2x + 1)$
   (i) $-3m(-2 + 4m + 3n)$

3. Remove the brackets and simplify the
   following:
   (a) $18 - 13(x + 2)$ (b) $x(x + y)$

4. Remove the brackets and simplify the
   following expressions:
   (a) $(x + 1)(x + 6)$ (b) $(x + 4)(x + 5)$
   (c) $(x - 2)(x + 3)$ (d) $(x + 6)(x - 1)$
   (e) $(x + y)(m + n)$ (f) $(4 + y)(3 + x)$
   (g) $(5 - x)(5 + x)$
   (h) $(17x + 2)(3x - 5)$

5. Remove the brackets and simplify the following expressions:
    (a) $(x+3)(x-7)$ (b) $(2x-1)(3x+7)$
    (c) $(4x+1)(4x-1)$
    (d) $(x+3)(x-3)$ (e) $(2-x)(3+2x)$

6. Remove the brackets and simplify the following expressions:
    (a) $\dfrac{1}{2}(x+2y)+\dfrac{7}{2}(4x-y)$
    (b) $\dfrac{3}{4}(x-1)+\dfrac{1}{4}(2x+8)$

7. Remove the brackets from
    (a) $-(x-y)$ (b) $-(a+2b)$
    (c) $-\dfrac{1}{2}(3p+q)$

8. Remove the brackets from $(x+1)(x+2)$. Use your result to remove the brackets from $(x+1)(x+2)(x+3)$.

Test and assignment exercises 7

1. Simplify
    (a) $7x^2+4x^2+9x-8x$ (b) $y+7-18y+1$ (c) $a^2+b^2+a^3-3b^2$

2. Simplify
    (a) $(3a^2b)\times(-a^3b^2c)$ (b) $\dfrac{x^3}{-x^2}$

3. Remove the brackets from
    (a) $(a+3b)(7a-2b)$ (b) $x^2(x+2y)$ (c) $x(x+y)(x-y)$

4. Remove the brackets from
    (a) $(7x+2)(3x-1)$ (b) $(1-x)(x+3)$ (c) $(5+x)x$ (d) $(8x+4)(7x-2)$

5. Remove the brackets and simplify
    (a) $3x(x+2)-7x^2$ (b) $-(2a+3b)(a+b)$ (c) $4(x+7)+13(x-2)$
    (d) $5(2a+5)-3(5a-2)$ (e) $\dfrac{1}{2}(a+4b)+\dfrac{3}{2}a$

Factorisation

8

Objectives: This chapter:

- explains what is meant by the 'factors' of an algebraic expression
- shows how an algebraic expression can be factorised
- shows how to factorise quadratic expressions

8.1 Factors and common factors

Recall from Chapter 1 that a number is **factorised** when it is written as a product. For example, 15 may be factorised into 3×5. We say that 3 and 5 are **factors** of 15. The number 16 can be written as 8×2 or 4×4, or even as 16×1, and so the factorisation may not be unique.

Algebraic expressions can also be factorised. Consider the expression $5x + 20y$. Both $5x$ and $20y$ have the number 5 common to both terms. We say that 5 is a **common factor**. Any common factors can be written outside a bracketed term. Thus $5x + 20y = 5(x + 4y)$. Removal of the brackets will result in the original expression and can always be used to check your answer. We see that factorisation can be thought of as reversing the process of removing brackets. Similarly, if we consider the expression $x^2 + 2x$, we note that both terms contain the factor x, and so $x^2 + 2x$ can be written as $x(x + 2)$. Hence x and $x + 2$ are both factors of $x^2 + 2x$.

WORKED EXAMPLES

8.1 Factorise $3x + 12$.

Solution The number 12 can be factorised as 3×4 so that 3 is a common factor of $3x$ and 12. We can write $3x + 12 = 3x + 3(4)$. Any common factors are written in front of the brackets and the contents of the brackets are

adjusted accordingly. So

$$3x + 3(4) = 3(x + 4)$$

Note again that this answer can be checked by removing the brackets.

8.2 List the ways in which $15x^2$ can be written as a product of its factors.

Solution $15x^2$ can be written in many different ways. Some of these are $15x^2 \times 1$, $15x \times x$, $15 \times x^2$, $5x \times 3x$, $5 \times 3x^2$ and $3 \times 5x^2$.

8.3 Factorise $8x^2 - 12x$.

Solution We can write $8x^2 - 12x = (4x)(2x) - (4x)3$ so that both terms contain the factor $4x$. This is placed at the front of the brackets to give

$$8x^2 - 12x = 4x(2x - 3)$$

8.4 What factors are common to the terms $5x^2$ and $15x^3$? Factorise $5x^2 + 15x^3$.

Solution Both terms contain a factor of 5. Because x^3 can be written as $x^2 \times x$, both $5x^2$ and $15x^3$ contain a factor x^2. Therefore

$$5x^2 + 15x^3 = 5x^2 + (5x^2)(3x) = 5x^2(1 + 3x)$$

8.5 Factorise $6x + 3x^2 + 9xy$.

Solution By careful inspection of all of the terms we see that $3x$ is a factor of each term. Hence

$$6x + 3x^2 + 9xy = 3x(2 + x + 3y)$$

Hence the factors of $6x + 3x^2 + 9xy$ are $3x$ and $2 + x + 3y$.

Self-assessment question 8.1

1. Explain what is meant by 'factorising an expression'.

Exercise 8.1

1. Remove the brackets from
   (a) $9(x + 3)$ (b) $-5(x - 2)$
   (c) $\dfrac{1}{2}(x + 1)$
   (d) $-(a - 3b)$ (e) $\dfrac{1}{2(x + y)}$
   (f) $\dfrac{x}{y(x - y)}$

2. List all the factors of each of (a) $4x^2$, (b) $6x^3$.

3. Factorise
   (a) $3x + 18$ (b) $3y - 9$ (c) $-3y - 9$
   (d) $-3 - 9y$ (e) $20 + 5t$ (f) $20 - 5t$
   (g) $-5t - 20$ (h) $3x + 12$ (i) $17t + 34$
   (j) $-36 + 4t$

4. Factorise
   (a) $x^4 + 2x$ (b) $x^4 - 2x$ (c) $3x^4 - 2x$
   (d) $3x^4 + 2x$ (e) $3x^4 + 2x^2$
   (f) $3x^4 + 2x^3$ (g) $17z - z^2$
   (h) $-xy + 3x$ (i) $-xy + 3y$
   (j) $x + 2xy + 3xyz$

5. Factorise
   (a) $10x + 20y$ (b) $12a + 3b$
   (c) $4x - 6xy$ (d) $7a + 14$

   (e) $10m - 15$ (f) $\dfrac{1}{5a + 35b}$

   (g) $\dfrac{1}{5a^2 + 35ab}$

6. Factorise
   (a) $15x^2 + 3x$ (b) $4x^2 - 3x$
   (c) $4x^2 - 8x$ (d) $15 - 3x^2$
   (e) $10x^3 + 5x^2 + 15x^2y$
   (f) $6a^2b - 12ab^2$
   (g) $16abc - 8ab^2 + 24bc$

8.2 Factorising quadratic expressions

Expressions of the form $ax^2 + bx + c$, where a, b and c are numbers, are called **quadratic expressions**. The numbers b or c may equal zero but a must not be zero. The number a is called the **coefficient** of x^2, b is the coefficient of x, and c is called the **constant term**.
   We see that

$$2x^2 + 3x - 1, \quad x^2 + 3x + 2, \quad x^2 + 7 \quad \text{and} \quad 2x^2 - x$$

are all quadratic expressions.

Key point

An expression of the form $ax^2 + bx + c$, where a, b and c are numbers, is called a quadratic expression. The coefficient of x^2 is a, the coefficient of x is b, and the constant term is c.

To factorise such an expression means to express it as a product of two terms. For example, removing the brackets from $(x + 6)(x - 3)$ gives $x^2 + 3x - 18$ (see Worked Example 7.22). Reversing the process, $x^2 + 3x - 18$ can be factorised to $(x + 6)(x - 3)$. Not all quadratic expressions can be factorised in this way. We shall now explore how such factorisation is attempted.

Quadratic expressions where the coefficient of x^2 is 1

Consider the expression $(x + m)(x + n)$. Removing the brackets we find

$$(x + m)(x + n) = (x + m)x + (x + m)n$$

$$= x^2 + mx + nx + mn$$

$$= x^2 + (m + n)x + mn$$

Note that the coefficient of the x term is the sum $m + n$ and the constant term is the product mn. Using this information several quadratic expressions can be factorised by careful inspection. For example, suppose we wish to factorise $x^2 + 5x + 6$. We know that $x^2 + (m + n)x + mn$ can be factorised to $(x + m)(x + n)$. We seek values of m and n so that

$$x^2 + 5x + 6 = x^2 + (m + n)x + mn$$

Comparing the coefficients of x on both sides we require

$$5 = m + n$$

Comparing the constant terms on both sides we require

$$6 = mn$$

By inspection we see that $m = 3$ and $n = 2$ have this property and so

$$x^2 + 5x + 6 = (x + 3)(x + 2)$$

Note that the answer can be easily checked by removing the brackets again.

WORKED EXAMPLES

8.6   Factorise the quadratic expression $x^2 + 8x + 12$.

VIDEO

Solution   The factorisation of $x^2 + 8x + 12$ will be of the form $(x + m)(x + n)$. This means that mn must equal 12 and $m + n$ must equal 8. The two numbers must therefore be 2 and 6. So

$$x^2 + 8x + 12 = (x + 2)(x + 6)$$

Note again that the answer can be checked by removing the brackets.

8.7   Factorise $x^2 + 10x + 25$.

Solution   We try to factorise in the form $(x + m)(x + n)$. We require $m + n$ to equal 10 and mn to equal 25. If $m = 5$ and $n = 5$ this requirement is met. Therefore $x^2 + 10x + 25 = (x + 5)(x + 5)$. It is usual practice to write this as $(x + 5)^2$.

8.8   Factorise $x^2 - 121$.

Solution   In this example the x term is missing. We still attempt to factorise as $(x + m)(x + n)$. We require $m + n$ to equal 0 and mn to equal -121. Some thought shows that if $m = 11$ and $n = -11$ this requirement is met. Therefore $x^2 - 121 = (x + 11)(x - 11)$.

8.9   Factorise $x^2 - 5x + 6$.

Solution   We try to factorise in the form $(x + m)(x + n)$. We require $m + n$ to equal -5 and mn to equal 6. By inspection we see that if $m = -3$ and $n = -2$ this requirement is met. Therefore $x^2 - 5x + 6 = (x - 3)(x - 2)$.

Quadratic expressions where the coefficient of x^2 is not 1

These expressions are a little harder to factorise. All possible factors of the first and last terms must be found, and various combinations of these should be attempted until the required answer is found. This involves trial and error along with educated guesswork and practice.

WORKED EXAMPLES

8.10 Factorise, if possible, the expression $2x^2 + 11x + 12$.

Solution The factors of the first term, $2x^2$, are $2x$ and x. The factors of the last term, 12, are

$$12, 1 \quad -12, -1 \quad 6, 2 \quad -6, -2 \quad \text{and} \quad 4, 3 \quad -4, -3$$

We can try each of these combinations in turn to find which gives us a coefficient of x of 11. For example, removing the brackets from

$$(2x + 12)(x + 1)$$

gives

$$(2x + 12)(x + 1) = (2x + 12)x + (2x + 12)(1)$$
$$= 2x^2 + 12x + 2x + 12$$
$$= 2x^2 + 14x + 12$$

which has an incorrect middle term. By trying further combinations it turns out that the only one producing a middle term of $11x$ is $(2x + 3)(x + 4)$ because

$$(2x + 3)(x + 4) = (2x + 3)(x) + (2x + 3)(4)$$
$$= 2x^2 + 3x + 8x + 12$$
$$= 2x^2 + 11x + 12$$

so that $(2x + 3)(x + 4)$ is the correct factorisation.

8.11 Factorise, if possible, $4x^2 + 6x + 2$.

Solution Before we try to factorise this quadratic expression notice that there is a factor of 2 in each term so that we can write it as $2(2x^2 + 3x + 1)$. Now consider the quadratic expression $2x^2 + 3x + 1$. The factors of the first term, $2x^2$, are $2x$ and x. The factors of the last term, 1, are simply 1 and 1, or -1 and -1. We can try these combinations in turn to find which gives us a middle term of $3x$. Removing the brackets from $(2x + 1)(x + 1)$ gives $2x^2 + 3x + 1$, which has the correct middle term. Finally, we can write

$$4x^2 + 6x + 2 = 2(2x^2 + 3x + 1) = 2(2x + 1)(x + 1)$$

8.12 Factorise $6x^2 + 7x - 3$.

VIDEO

Solution   The first term may be factorised as $6x \times x$ and also as $3x \times 2x$. The factors of the last term are

$$3, -1 \quad \text{and} \quad -3, 1$$

We need to try each combination in turn to find which gives us a coefficient of x of 7. For example, removing the brackets from

$$(6x + 3)(x - 1)$$

gives

$$(6x + 3)(x - 1) = (6x + 3)x + (6x + 3)(-1)$$
$$= 6x^2 + 3x - 6x - 3$$
$$= 6x^2 - 3x - 3$$

which has an incorrect middle term. By trying further combinations it turns out that the only one producing a middle term of $7x$ is $(3x - 1)(2x + 3)$ because

$$(3x - 1)(2x + 3) = (3x - 1)2x + (3x - 1)(3)$$
$$= 6x^2 - 2x + 9x - 3$$
$$= 6x^2 + 7x - 3$$

The correct factorisation is therefore $(3x - 1)(2x + 3)$.

Until you have sufficient experience at factorising quadratic expressions you must be prepared to go through the process of trying all possible combinations until the correct answer is found.

Self-assessment question 8.2

1. Not all quadratic expressions can be factorised. Try to find an example of one such expression.

Exercise 8.2

1. Factorise the following quadratic expressions:
   (a) $x^2 + 3x + 2$ (b) $x^2 + 13x + 42$
   (c) $x^2 + 2x - 15$ (d) $x^2 + 9x - 10$
   (e) $x^2 - 11x + 24$ (f) $x^2 - 100$
   (g) $x^2 + 4x + 4$ (h) $x^2 - 36$
   (i) $x^2 - 25$ (j) $x^2 + 10x + 9$
   (k) $x^2 + 8x - 9$ (l) $x^2 - 8x - 9$
   (m) $x^2 - 10x + 9$ (n) $x^2 - 5x$

2. Factorise the following quadratic expressions:
   (a) $2x^2 - 5x - 3$ (b) $3x^2 - 5x - 2$
   (c) $10x^2 + 11x + 3$ (d) $2x^2 + 12x + 16$
   (e) $2x^2 + 5x + 3$ (f) $3s^2 + 5s + 2$
   (g) $3z^2 + 17z + 10$ (h) $9x^2 - 36$
   (i) $4x^2 - 25$

3. (a) By removing the brackets show that

    $$(x+y)(x-y) = x^2 - y^2$$

    This result is known as the **difference of two squares**.

    (b) Using the result in part (a) write down the factorisation of
    (i) $16x^2 - 1$ (ii) $16x^2 - 9$
    (iii) $25t^2 - 16r^2$

4. Factorise the following quadratic expressions:
    (a) $x^2 + 3x - 10$ (b) $2x^2 - 3x - 20$
    (c) $9x^2 - 1$ (d) $10x^2 + 14x - 12$
    (e) $x^2 + 15x + 26$ (f) $-x^2 - 2x + 3$

5. Factorise
    (a) $100 - 49x^2$ (b) $36x^2 - 25y^2$
    (c) $\frac{1}{4} - 9v^2$ (d) $\frac{x^2}{y^2} - 4$

Test and assignment exercises 8

1. Factorise the following expressions:
    (a) $7x + 49$ (b) $121x + 22y$ (c) $a^2 + ab$ (d) $ab + b^2$ (e) $ab^2 + ba^2$

2. Factorise the following quadratic expressions:
    (a) $3x^2 + x - 2$ (b) $x^2 - 144$ (c) $s^2 - 5s + 6$ (d) $2y^2 - y - 15$

3. Factorise the following:
    (a) $1 - x^2$ (b) $x^2 - 1$ (c) $9 - x^2$ (d) $x^2 - 81$ (e) $25 - y^2$

4. Factorise the denominators of the following expressions:
    (a) $\dfrac{1}{x^2 + 6x}$ (b) $\dfrac{3}{s^2 + 3s + 2}$ (c) $\dfrac{3}{s^2 + s - 2}$ (d) $\dfrac{5}{x^2 + 11x + 28}$ (e) $\dfrac{x}{2x^2 - 17x - 9}$

Algebraic fractions

9

Objectives: This chapter:

- explains how to simplify algebraic fractions by cancelling common factors
- explains how algebraic fractions can be multiplied and divided
- explains how algebraic fractions can be added and subtracted
- explains how to express a fraction as the sum of its partial fractions

9.1 Introduction

Just as one whole number divided by another is a numerical fraction, so one algebraic expression divided by another is called an **algebraic fraction**.

$$\frac{x}{y} \qquad \frac{x^2 + y}{x} \qquad \frac{3x + 2}{7}$$

are all examples of algebraic fractions. The top line is known as the **numerator** of the fraction, and the bottom line is the **denominator**.

Rules for determining the sign of the answer when dividing positive and negative algebraic expressions are the same as those used for dividing numbers.

Key point

When dividing

$$\frac{\text{positive}}{\text{positive}} = \text{positive} \qquad \frac{\text{negative}}{\text{positive}} = \text{negative}$$

$$\frac{\text{positive}}{\text{negative}} = \text{negative} \qquad \frac{\text{negative}}{\text{negative}} = \text{positive}$$

Using these rules we see that an algebraic expression can often be written in different but equivalent forms. For example, note that

$$\frac{x}{-y} \text{ can be written as } -\frac{x}{y}$$

and that

$$\frac{-x}{y} \text{ can be written as } -\frac{x}{y}$$

and also that

$$\frac{-x}{-y} \text{ can be written as } \frac{x}{y}$$

9.2 Cancelling common factors

Cancellation of common factors was described in detail in §2.2.

Consider the numerical fraction $\frac{3}{12}$. To simplify this we factorise both the numerator and the denominator. Any factors which appear in both the numerator and the denominator are called **common factors**. These can be cancelled. For example,

$$\frac{3}{12} = \frac{1 \times 3}{4 \times 3} = \frac{1 \times \cancel{3}}{4 \times \cancel{3}} = \frac{1}{4}$$

The same process is applied when dealing with algebraic fractions.

WORKED EXAMPLES

9.1   For each pair of expressions, state which factors are common to both.

(a) $3xy$ and $6xz$ (b) xy and $5y^2$ (c) $3(x+2)$ and $(x+2)^2$

(d) $3(x-1)$ and $(x-1)(x+4)$

Solution   (a) The expression $6xz$ can be written $(3)(2)xz$. We see that factors common to both this and $3xy$ are 3 and x.

(b) The expression $5y^2$ can be written $5(y)(y)$. We see that the only factor common to both this and xy is y.

(c) $(x+2)^2$ can be written $(x+2)(x+2)$. Thus $(x+2)$ is a factor common to both $(x+2)^2$ and $3(x+2)$.

(d) $3(x-1)$ and $(x-1)(x+4)$ have a common factor of $(x-1)$.

9.2   Simplify

$$\frac{18x^2}{6x}$$

Solution   First note that 18 can be factorised as 6×3. So there are factors of 6 and x in both the numerator and the denominator. Then common factors can be cancelled. That is,

$$\frac{18x^2}{6x} = \frac{(6)(3)x^2}{6x} = \frac{3x}{1} = 3x$$

Key point   When simplifying an algebraic fraction only factors common to both the numerator and denominator can be cancelled.

A fraction is expressed in its simplest form by factorising the numerator and denominator and cancelling any common factors.

WORKED EXAMPLES

9.3   Simplify

$$\frac{5}{25 + 15x}$$

Solution   First of all note that the denominator can be factorised as $5(5 + 3x)$. There is therefore a factor of 5 in both the numerator and denominator. So 5 is a common factor. This can be cancelled. That is,

$$\frac{5}{25 + 15x} = \frac{1 \times 5}{5(5 + 3x)} = \frac{1 \times \cancel{5}}{\cancel{5}(5 + 3x)} = \frac{1}{5 + 3x}$$

It is very important to note that the number 5 that has been cancelled is a common factor. It is incorrect to try to cancel terms that are not common factors.

9.4   Simplify

VIDEO

$$\frac{5x}{25x + 10y}$$

Solution   Factorising the denominator we can write

$$\frac{5x}{25x + 10y} = \frac{5x}{5(5x + 2y)}$$

We see that there is a common factor of 5 in both numerator and denominator that can be cancelled. Thus

$$\frac{5x}{25x + 10y} = \frac{\cancel{5}x}{\cancel{5}(5x + 2y)} = \frac{x}{5x + 2y}$$

Note that no further cancellation is possible. x is not a common factor because it is not a factor of the denominator.

9.5 Simplify

$$\frac{4x}{3x^2 + x}$$

Solution Note that the denominator factorises to $x(3x + 1)$. Once both numerator and denominator have been factorised, any common factors are cancelled. So

$$\frac{4x}{3x^2 + x} = \frac{4x}{x(3x + 1)} = \frac{4\cancel{x}}{\cancel{x}(3x + 1)} = \frac{4}{3x + 1}$$

Note that the factor x is common to both numerator and denominator and so has been cancelled.

9.6 Simplify

$$\frac{x}{x^2 + 2x}$$

Solution Note that the denominator factorises to $x(x + 2)$. Also note that the numerator can be written as $1 \times x$. So

$$\frac{x}{x^2 + 2x} = \frac{1 \times x}{x(x + 2)} = \frac{1 \times \cancel{x}}{\cancel{x}(x + 2)} = \frac{1}{x + 2}$$

9.7 Simplify

(a) $\dfrac{2(x - 1)}{(x + 3)(x - 1)}$ (b) $\dfrac{x - 4}{(x - 4)^2}$

Solution (a) There is a factor of $(x - 1)$ common to both the numerator and denominator. This is cancelled to give

$$\frac{2(x - 1)}{(x + 3)(x - 1)} = \frac{2}{x + 3}$$

(b) There is a factor of $x - 4$ in both numerator and denominator. This is cancelled as follows:

$$\frac{x - 4}{(x - 4)^2} = \frac{1(x - 4)}{(x - 4)(x - 4)} = \frac{1}{x - 4}$$

9.8 Simplify

VIDEO

$$\frac{x+2}{x^2+3x+2}$$

Solution The denominator is factorised and then any common factors are cancelled:

$$\frac{x+2}{x^2+3x+2}=\frac{1(x+2)}{(x+2)(x+1)}=\frac{1}{x+1}$$

9.9 Simplify

(a) $\dfrac{3x+xy}{x^2+5x}$ (b) $\dfrac{x^2-1}{x^2+3x+2}$

Solution The numerator and denominator are both factorised and any common factors are cancelled:

(a) $\dfrac{3x+xy}{x^2+5x}=\dfrac{x(3+y)}{x^2+5x}=\dfrac{\cancel{x}(3+y)}{\cancel{x}(x+5)}=\dfrac{3+y}{5+x}$

(b) $\dfrac{x^2-1}{x^2+3x+2}=\dfrac{(x+1)(x-1)}{(x+1)(x+2)}=\dfrac{x-1}{x+2}$

Self-assessment questions 9.2

1. Explain why no cancellation is possible in the expression $\dfrac{3x}{3x+y}$.

2. Explain why no cancellation is possible in the expression $\dfrac{x+1}{x+3}$.

3. Explain why it is possible to perform a cancellation in the expression $\dfrac{x+1}{2x+2}$, and perform it.

Exercise 9.2

MyMathLab

1. Simplify
(a) $\dfrac{9x}{3y}$ (b) $\dfrac{9x}{x^2}$ (c) $\dfrac{9xy}{3x}$ (d) $\dfrac{9xy}{3y}$
(e) $\dfrac{9xy}{xy}$ (f) $\dfrac{9xy}{3xy}$

2. Simplify
(a) $\dfrac{15x}{3y}$ (b) $\dfrac{15x}{5y}$ (c) $\dfrac{15xy}{x}$ (d) $\dfrac{15xy}{xy}$
(e) $\dfrac{x^5}{-x^3}$ (f) $\dfrac{-y^3}{y^7}$ (g) $\dfrac{-y}{-y^2}$ (h) $\dfrac{-y^{-3}}{-y^4}$

3. Simplify the following algebraic fractions:
(a) $\dfrac{4}{12+8x}$ (b) $\dfrac{5+10x}{5}$ (c) $\dfrac{2}{4+14x}$
(d) $\dfrac{2x}{4+14x}$ (e) $\dfrac{2x}{2+14x}$ (f) $\dfrac{7}{49x+7y}$
(g) $\dfrac{7y}{49x+7y}$ (h) $\dfrac{7x}{49x+7y}$

4. Simplify

   (a) $\dfrac{15x + 3}{3}$ (b) $\dfrac{15x + 3}{3x + 6y}$ (c) $\dfrac{12}{4x + 8}$

   (d) $\dfrac{12x}{4xy + 8x}$ (e) $\dfrac{13x}{x^2 + 5x}$

   (f) $\dfrac{17y}{9y^2 + 4y}$

5. Simplify the following:

   (a) $\dfrac{5}{15 + 10x}$ (b) $\dfrac{2x}{x^2 + 7x}$

   (c) $\dfrac{2x + 8}{x^2 + 2x - 8}$ (d) $\dfrac{7ab}{a^2b^2 + 9ab}$

   (e) $\dfrac{xy}{xy + x}$

6. Simplify

   (a) $\dfrac{x - 4}{(x - 4)(x - 2)}$ (b) $\dfrac{2x - 4}{x^2 + x - 6}$

   (c) $\dfrac{3x}{3x^2 + 6x}$ (d) $\dfrac{x^2 + 2x + 1}{x^2 - 2x - 3}$

   (e) $\dfrac{2(x - 3)}{(x - 3)^2}$ (f) $\dfrac{x - 3}{(x - 3)^2}$

   (g) $\dfrac{x - 3}{2(x - 3)^2}$ (h) $\dfrac{4(x - 3)}{2(x - 3)^2}$

   (i) $\dfrac{x + 4}{2(x + 4)^2}$ (j) $\dfrac{x + 4}{2(x + 4)}$

   (k) $\dfrac{2(x + 4)}{(x + 4)}$ (l) $\dfrac{(x + 4)(x - 3)}{x - 3}$

   (m) $\dfrac{x + 4}{(x - 3)(x + 4)}$ (n) $\dfrac{x + 3}{x^2 + 7x + 12}$

   (o) $\dfrac{x + 4}{2x + 8}$ (p) $\dfrac{x + 4}{2x + 9}$

9.3 Multiplication and division of algebraic fractions

To multiply two algebraic fractions together we multiply their numerators together and multiply their denominators together:

Key point

$$\frac{a}{b} \times \frac{c}{d} = \frac{a \times c}{b \times d}$$

Any common factors in the result should be cancelled.

WORKED EXAMPLES

9.10   Simplify

$$\frac{4}{5} \times \frac{x}{y}$$

Solution   We multiply the numerators together and multiply the denominators together. That is,

$$\frac{4}{5} \times \frac{x}{y} = \frac{4x}{5y}$$

9.11   Simplify

$$\frac{4}{x} \times \frac{3y}{16}$$

Solution   The numerators are multiplied together and the denominators are multiplied together. Therefore

$$\frac{4}{x} \times \frac{3y}{16} = \frac{4 \times 3y}{16x}$$

Because $16x = 4 \times 4x$, the common factor 4 can be cancelled. So

$$\frac{4 \times 3y}{16x} = \frac{\cancel{4} \times 3y}{\cancel{4} \times 4x} = \frac{3y}{4x}$$

9.12   Simplify

(a) $\dfrac{1}{2} \times x$ (b) $\dfrac{1}{2} \times (a + b)$

Solution   (a) Writing x as $\dfrac{x}{1}$ we can state

$$\frac{1}{2} \times x = \frac{1}{2} \times \frac{x}{1} = \frac{1 \times x}{2 \times 1} = \frac{x}{2}$$

(b) Writing $a + b$ as $\dfrac{a + b}{1}$ we can state

$$\frac{1}{2} \times (a + b) = \frac{1}{2} \times \frac{(a + b)}{1} = \frac{1 \times (a + b)}{2 \times 1} = \frac{a + b}{2}$$

9.13   Simplify

$$\frac{4x^2}{y} \times \frac{3x^3}{yz}$$

Solution   We multiply the numerators together and multiply the denominators together:

$$\frac{4x^2}{y} \times \frac{3x^3}{yz} = \frac{4x^2 \times 3x^3}{y \times yz} = \frac{12x^5}{y^2z}$$

9.14   Simplify

$$5 \times \left(\frac{x - 3}{25} \right)$$

Solution   This means

$$\frac{5}{1} \times \frac{x - 3}{25}$$

which equals

$$\frac{5 \times (x - 3)}{1 \times 25}$$

A common factor of 5 can be cancelled from the numerator and denominator to give

$$\frac{(x - 3)}{5}$$

9.15   Simplify

$$-\frac{1}{5} \times \frac{3x - 4}{8}$$

Solution   We can write

$$-\frac{1}{5} \times \frac{3x - 4}{8} = -\frac{1 \times (3x - 4)}{5 \times 8} = -\frac{3x - 4}{40}$$

Note that the answer can also be expressed as $\frac{(4 - 3x)}{40}$ because

$$-\frac{3x - 4}{40} = \frac{-1}{1} \times \frac{3x - 4}{40} = \frac{-3x + 4}{40} = \frac{4 - 3x}{40}$$

You should be aware from the last worked example that a solution can often be expressed in a number of equivalent ways.

WORKED EXAMPLES

9.16   Simplify

$$\frac{a}{a + b} \times \frac{b}{5a^2}$$

Solution

$$\frac{a}{a+b} \times \frac{b}{5a^2} = \frac{ab}{5a^2(a+b)}$$

Cancelling the common factor of a in numerator and denominator gives

$$\frac{b}{5a(a+b)}$$

9.17   Simplify

VIDEO

$$\frac{x^2+4x+3}{2x+8} \times \frac{x+4}{x+1}$$

Solution   Before multiplying the two fractions together we should try to factorise if possible so that common factors can be identified. By factorising, we can write the given expressions as

$$\frac{(x+1)(x+3)}{2(x+4)} \times \frac{x+4}{x+1} = \frac{(x+1)(x+3)(x+4)}{2(x+4)(x+1)}$$

Cancelling common factors this simplifies to just

$$\frac{x+3}{2}$$

Division is performed by inverting the second fraction and multiplying:

Key point

$$\frac{a}{b} \div \frac{c}{d} = \frac{a}{b} \times \frac{d}{c}$$

WORKED EXAMPLES

9.18   Simplify

$$\frac{10a}{b} \div \frac{a^2}{3b}$$

Solution   The second fraction is inverted and then multiplied by the first. That is,

$$\frac{10a}{b} \div \frac{a^2}{3b} = \frac{10a}{b} \times \frac{3b}{a^2} = \frac{30ab}{a^2b} = \frac{30}{a}$$

9.19   Simplify

$$\frac{x^2y^3}{z} \div \frac{y}{x}$$

Solution $\dfrac{x^2y^3}{z} \div \dfrac{y}{x} = \dfrac{x^2y^3}{z} \times \dfrac{x}{y} = \dfrac{x^3y^3}{zy}$

Any common factors in the result can be cancelled. So

$$\dfrac{x^3y^3}{zy} = \dfrac{x^3y^2}{z}$$

Self-assessment question 9.3

1. The technique of multiplying and dividing algebraic fractions is identical to that used for numbers. True or false?

Exercise 9.3

1. Simplify

   (a) $\dfrac{1}{2} \times \dfrac{y}{3}$ (b) $\dfrac{1}{3} \times \dfrac{z}{2}$ (c) $\dfrac{2}{5}$ of $\dfrac{1}{y}$

   (d) $\dfrac{2}{5}$ of $\dfrac{1}{x}$ (e) $\dfrac{3}{4}$ of $\dfrac{x}{y}$ (f) $\dfrac{3}{5} \times \dfrac{x^2}{y}$

   (g) $\dfrac{x}{y} \times \dfrac{3}{5}$ (h) $\dfrac{7}{8} \times \dfrac{x}{2y}$ (i) $\dfrac{1}{2} \times \dfrac{1}{2x}$

   (j) $\dfrac{1}{2} \times \dfrac{x}{2}$ (k) $\dfrac{1}{2} \times \dfrac{2}{x}$ (l) $\dfrac{1}{3} \times \dfrac{x}{3}$

   (m) $\dfrac{1}{3} \times \dfrac{3}{x}$ (n) $\dfrac{1}{3} \times \dfrac{1}{3x}$ (o) $\dfrac{1}{3} \times \dfrac{3x}{2}$

2. Simplify

   (a) $\dfrac{1}{2} \div \dfrac{x}{2}$ (b) $\dfrac{1}{2} \div \dfrac{2}{x}$ (c) $\dfrac{2}{x} \div \dfrac{2}{x}$

   (d) $\dfrac{x}{2} \div \dfrac{1}{2}$ (e) $\dfrac{2}{x} \div 2$ (f) $\dfrac{2}{x} \div \dfrac{1}{2}$

   (g) $\dfrac{3}{x} \div \dfrac{1}{2}$

3. Simplify the following:

   (a) $\dfrac{5}{4} \times \dfrac{a}{25}$ (b) $\dfrac{5}{4} \times \dfrac{a}{b}$ (c) $\dfrac{8a}{b^2} \times \dfrac{b}{16a^2}$

   (d) $\dfrac{9x}{3y} \times \dfrac{2x}{y^2}$ (e) $\dfrac{3}{5a} \times \dfrac{b}{a}$ (f) $\dfrac{1}{4} \times \dfrac{x}{y}$

   (g) $\dfrac{1}{3} \times \dfrac{x}{x+y}$ (h) $\dfrac{x-3}{x+4} \times \dfrac{1}{3x-9}$

VIDEO

4. Simplify the following:

   (a) $\dfrac{3}{x} \times \dfrac{xy}{z^3}$ (b) $\dfrac{(3+x)}{x} \div \dfrac{y}{x}$ (c) $\dfrac{4}{3} \div \dfrac{16}{x}$

   (d) $\dfrac{a}{bc^2} \times \dfrac{b^2c}{a}$

5. Simplify

   (a) $\dfrac{x+2}{(x+5)(x+4)} \times \dfrac{x+5}{x+2}$

   (b) $\dfrac{x-2}{4} \div \dfrac{x}{16}$ (c) $\dfrac{12ab}{5ef} \div \dfrac{4ab^2}{f}$

   (d) $\dfrac{x+3y}{2x} \div \dfrac{y}{4x^2}$ (e) $\dfrac{3}{x} \times \dfrac{3}{y} \times \dfrac{1}{z}$

6. Simplify

   $$\dfrac{1}{x+1} \times \dfrac{2x+2}{x+3}$$

7. Simplify

   $$\dfrac{x+1}{x+2} \times \dfrac{x^2+6x+8}{x^2+4x+3}$$

9.4 Addition and subtraction of algebraic fractions

For revision of adding and subtracting fractions see §2.3.

The method is the same as that for adding or subtracting numerical fractions. Note that it is not correct simply to add or subtract the numerator and denominator. The lowest common denominator must first be found. This is the simplest expression that contains all original denominators as its factors. Each fraction is then written with this common denominator. The fractions can then be added or subtracted by adding or subtracting just the numerators, and dividing the result by the common denominator.

WORKED EXAMPLES

9.20 Add the fractions $\frac{3}{4}$ and $\frac{1}{x}$.

Solution We must find $\frac{3}{4} + \frac{1}{x}$. To do this we must first rewrite the fractions to ensure they have a common denominator. The common denominator is the simplest expression that has the given denominators as its factors. The simplest such expression is $4x$. We write

$$\frac{3}{4} \text{ as } \frac{3x}{4x} \qquad \text{and} \qquad \frac{1}{x} \text{ as } \frac{4}{4x}$$

Then

$$\frac{3}{4} + \frac{1}{x} = \frac{3x}{4x} + \frac{4}{4x}$$

$$= \frac{3x + 4}{4x}$$

No further simplification is possible.

9.21 Simplify

$$\frac{3}{x} + \frac{4}{x^2}$$

Solution The expression $\frac{3}{x}$ is rewritten as $\frac{3x}{x^2}$, which makes the denominators of both terms x^2 but leaves the value of the expression unaltered. Note that both the original denominators, x and x^2, are factors of the new denominator. We call this denominator the **lowest common denominator**. The fractions are then added by adding just the numerators. That is,

$$\frac{3x}{x^2} + \frac{4}{x^2} = \frac{3x + 4}{x^2}$$

9.22   Express $\dfrac{5}{a} - \dfrac{4}{b}$ as a single fraction.

Solution   Both fractions are rewritten to have the same denominator. The simplest expression containing both a and b as its factors is ab. Therefore ab is the lowest common denominator. Then

$$\frac{5}{a} - \frac{4}{b} = \frac{5b}{ab} - \frac{4a}{ab} = \frac{5b - 4a}{ab}$$

9.23   Write $\dfrac{4}{x+y} - \dfrac{3}{y}$ as a single fraction.

Solution   The simplest expression that contains both denominators as its factors is $(x+y)y$. We must rewrite each term so that it has this denominator:

$$\frac{4}{x+y} = \frac{4}{x+y} \times \frac{y}{y} = \frac{4y}{(x+y)y}$$

Similarly,

$$\frac{3}{y} = \frac{3}{y} \times \frac{x+y}{x+y} = \frac{3(x+y)}{(x+y)y}$$

The fractions are then subtracted by subtracting just the numerators:

$$\frac{4}{x+y} - \frac{3}{y} = \frac{4y}{(x+y)y} - \frac{3(x+y)}{(x+y)y} = \frac{4y - 3(x+y)}{(x+y)y}$$

which simplifies to

$$\frac{y - 3x}{(x+y)y}$$

9.24   Express as a single fraction

VIDEO

$$\frac{2}{x+3} + \frac{5}{x-1}$$

Solution   The simplest expression having both $x+3$ and $x-1$ as its factors is

$$(x+3)(x-1)$$

This is the lowest common denominator. Each term is rewritten so that it has this denominator. Thus

$$\frac{2}{x+3} = \frac{2(x-1)}{(x+3)(x-1)} \qquad \text{and} \qquad \frac{5}{x-1} = \frac{5(x+3)}{(x+3)(x-1)}$$

Then

$$\frac{2}{x+3}+\frac{5}{x-1}=\frac{2(x-1)}{(x+3)(x-1)}+\frac{5(x+3)}{(x+3)(x-1)}$$

$$=\frac{2(x-1)+5(x+3)}{(x+3)(x-1)}$$

which simplifies to

$$\frac{7x+13}{(x+3)(x-1)}$$

9.25   Express as a single fraction

VIDEO

$$\frac{1}{x-4}+\frac{1}{(x-4)^2}$$

Solution   The simplest expression having $x-4$ and $(x-4)^2$ as its factors is $(x-4)^2$. Both fractions are rewritten with this denominator:

$$\frac{1}{x-4}+\frac{1}{(x-4)^2}=\frac{(x-4)}{(x-4)^2}+\frac{1}{(x-4)^2}$$

$$=\frac{x-4+1}{(x-4)^2}$$

$$=\frac{x-3}{(x-4)^2}$$

Self-assessment question 9.4

1. Explain what is meant by the 'lowest common denominator' and how it is found.

Exercise 9.4

1. Express each of the following as a single fraction:

    (a) $\dfrac{z}{2}+\dfrac{z}{3}$ (b) $\dfrac{x}{3}+\dfrac{x}{4}$ (c) $\dfrac{y}{5}+\dfrac{y}{25}$

2. Express each of the following as a single fraction:

    (a) $\dfrac{1}{2}+\dfrac{1}{x}$ (b) $\dfrac{1}{2}+x$ (c) $\dfrac{1}{3}+y$

    (d) $\dfrac{1}{3}+\dfrac{1}{y}$ (e) $8+\dfrac{1}{y}$

3. Express each of the following as a single fraction:

    (a) $\dfrac{5}{x}-\dfrac{1}{2}$ (b) $\dfrac{5}{x}+2$ (c) $\dfrac{3}{x}-\dfrac{1}{3}$

    (d) $\dfrac{x}{3}-\dfrac{1}{2}$ (e) $\dfrac{3}{x}+\dfrac{1}{3}$

4. Express each of the following as a single fraction:

(a) $\dfrac{3}{x}+\dfrac{4}{y}$ (b) $\dfrac{3}{x^2}+\dfrac{4y}{x}$ (c) $\dfrac{4ab}{x}+\dfrac{3ab}{2y}$

(d) $\dfrac{4xy}{a}+\dfrac{3xy}{2b}$ (e) $\dfrac{3}{x}-\dfrac{6}{2x}$ (f) $\dfrac{3x}{2y}-\dfrac{7y}{4x}$

(g) $\dfrac{3}{x+y}-\dfrac{2}{y}$ (h) $\dfrac{1}{a+b}-\dfrac{1}{a-b}$

(i) $2x+\dfrac{1}{2x}$ (j) $2x-\dfrac{1}{2x}$

5. Express each of the following as a single fraction:

(a) $\dfrac{x}{y}+\dfrac{3x^2}{z}$ (b) $\dfrac{4}{a}+\dfrac{5}{b}$

(c) $\dfrac{6x}{y}-\dfrac{2y}{x}$ (d) $3x-\dfrac{3x+1}{4}$

(e) $\dfrac{5a}{12}+\dfrac{9a}{18}$ (f) $\dfrac{x-3}{4}+\dfrac{3}{5}$

6. Express each of the following as a single fraction:

(a) $\dfrac{1}{x+1}+\dfrac{1}{x+2}$ (b) $\dfrac{1}{x-1}+\dfrac{2}{x+3}$

(c) $\dfrac{3}{x+5}+\dfrac{1}{x+4}$ (d) $\dfrac{1}{x-2}+\dfrac{3}{x-4}$

(e) $\dfrac{3}{2x+1}+\dfrac{1}{x+1}$ (f) $\dfrac{3}{1-2x}+\dfrac{1}{x}$

(g) $\dfrac{3}{x+1}+\dfrac{4}{(x+1)^2}$

(h) $\dfrac{1}{x-1}+\dfrac{1}{(x-1)^2}$

9.5 Partial fractions

We have seen how to add and/or subtract algebraic fractions to yield a single fraction. Section 9.4 illustrates the process with some examples.

Sometimes we wish to use the reverse of this process. That is, starting with a single fraction we wish to express it as the sum of two or more simpler fractions. Each of these simpler fractions is known as a **partial fraction** because it is part of the original fraction.

Let us refer to Worked Example 9.24. From this example we can see that

$$\frac{2}{x+3}+\frac{5}{x-1}$$

can be expressed as the single fraction, $\frac{7x+13}{(x+3)(x-1)}$. Worked Example 9.26 illustrates the process of starting with $\frac{7x+13}{(x+3)(x-1)}$ and finding its partial fractions, $\frac{2}{x+3}$ and $\frac{5}{x-1}$.

WORKED EXAMPLES

9.26 Express $\frac{7x+13}{(x+3)(x-1)}$ as its partial fractions.

Solution   The denominator has two factors: $x + 3$ and $x - 1$. It is these factors that determine the form of the partial fractions. Each factor in the denominator produces a partial fraction – this is an important point.

The factor $x + 3$ produces a partial fraction of the form $\frac{A}{x+3}$ where A is a constant. Similarly the factor $x - 1$ produces a partial fraction of the form $\frac{B}{x-1}$ where B is a constant. Hence we have

$$\frac{7x + 13}{(x + 3)(x - 1)} = \frac{A}{x + 3} + \frac{B}{x - 1} \tag{9.1}$$

We now need to find the values of A and B. By multiplying both sides of (9.1) by $(x + 3)(x - 1)$ we have

$$\frac{7x + 3}{(x + 3)(x - 1)} \times (x + 3)(x - 1) = \frac{A}{x + 3} \times (x + 3)(x - 1)$$

$$+ \frac{B}{x - 1} \times (x + 3)(x - 1) \tag{9.2}$$

By cancelling the common factors in each term, (9.2) simplifies to

$$7x + 13 = A(x - 1) + B(x + 3) \tag{9.3}$$

Note that (9.3) is true for *all* values of x. To find the values of A and B we can substitute into (9.3) any value of x we choose. We choose values of x that are most helpful and convenient. Let us choose x to be -3. When $x = -3$ is substituted into (9.3) we obtain

$$7(-3) + 13 = A(-3 - 1) + B(-3 + 3)$$

from which

$$-8 = A(-4)$$
$$A = 2$$

As you can see, the value of $x = -3$ was chosen in order to simplify (9.3) in such a way that the value of A could then be found.

Now we return to (9.3) and let $x = 1$. Then (9.3) simplifies to

$$7(1) + 13 = A(1 - 1) + B(1 + 3)$$

from which $B = 5$. Clearly we chose $x = 1$ in order to simplify (9.3) by eliminating the A term, thus allowing B to be found. Putting $A = 2$, $B = 5$ into (9.1) we have the partial fractions

$$\frac{7x + 13}{(x + 3)(x - 1)} = \frac{2}{x + 3} + \frac{5}{x - 1}$$

9.27   Express $\frac{5x+28}{x^2+7x+10}$ as partial fractions.

VIDEO

Solution   The factors of the denominator must first be found, that is $x^2 + 7x + 10$ must be factorised:

$$x^2 + 7x + 10 = (x + 2)(x + 5)$$

As the denominator has two factors, then there are two partial fractions. The factor $x + 2$ leads to a partial fraction $\frac{A}{x+2}$ and the factor $x + 5$ leads to a partial fraction $\frac{B}{x+5}$, where A and B are constants whose values have yet to be found.

So

$$\frac{5x + 28}{x^2 + 7x + 10} = \frac{5x + 28}{(x + 2)(x + 5)} = \frac{A}{x + 2} + \frac{B}{x + 5} \tag{9.4}$$

To find the values of A and B we multiply both sides of (9.4) by $(x + 2)(x + 5)$. After cancelling common factors in each term we have

$$5x + 28 = A(x + 5) + B(x + 2) \tag{9.5}$$

We now select convenient values of x to simplify (9.5) so that A and B can be found. We see that choosing $x = -2$ will simplify (9.5) so that A can be determined. With $x = -2$, (9.5) becomes

$$5(-2) + 28 = A(-2 + 5) + B(-2 + 2)$$
$$18 = 3A$$
$$A = 6$$

Returning to (9.5), we let $x = -5$ so that B can be found:

$$5(-5) + 28 = A(-5 + 5) + B(-5 + 2)$$
$$3 = -3B$$
$$B = -1$$

Putting the values of A and B into (9.4) yields the partial fractions

$$\frac{5x + 28}{x^2 + 7x + 10} = \frac{6}{x + 2} - \frac{1}{x + 5}$$

If a denominator has a repeated factor then a slight variation of the previous method is employed.

WORKED EXAMPLE

9.28   Find the partial fractions of $\frac{6x-5}{4x^2-4x+1}$

VIDEO

Solution   As in the previous example the denominator must first be factorised:

$$4x^2 - 4x + 1 = (2x - 1)(2x - 1) = (2x - 1)^2$$

We note that although there are two factors, they are identical (that is, a repeated factor). As there are two factors, then two partial fractions are generated. The partial fractions in such a case are of the form

$$\frac{A}{2x-1} + \frac{B}{(2x-1)^2}$$

where A and B are constants. Note that the denominators $(2x-1)$ and $(2x-1)^2$ are used. So

$$\frac{6x-5}{4x^2-4x+1} = \frac{6x-5}{(2x-1)^2} = \frac{A}{2x-1} + \frac{B}{(2x-1)^2} \qquad (9.6)$$

We need to find the values of the constants A and B. Multiplying both sides of (9.6) by $(2x-1)^2$ and then cancelling any common factors yields:

$$6x-5 = A(2x-1) + B \qquad (9.7)$$

The left-hand side and right-hand side of (9.7) are equal for all values of x. We choose convenient values of x to help us find the values of A and B. By substituting $x = \frac{1}{2}$ into (9.7) it simplifies to

$$-2 = A(0) + B$$

and so $B = -2$. The value of $x = \frac{1}{2}$ was chosen as the factor $(2x-1)$ then evaluated to 0 and the A term on the right-hand side became 0, allowing B to be found directly. We now need to find the value of A. It is impossible to eliminate the B term from (9.7) by any choice of an x value. Hence we simply choose any value of x that simplifies (9.7) considerably. Let us for example substitute $x = 0$ into (9.7) to obtain

$$-5 = A(-1) + B$$
$$A = B + 5$$

As we have already found the value of B to be -2, this is then substituted into the above equation to yield $A = 3$.

Substituting the values of A and B into (9.6) produces the partial fractions

$$\frac{6x-5}{4x^2-4x+1} = \frac{3}{2x-1} - \frac{2}{(2x-1)^2}$$

From Worked Examples 9.26, 9.27 and 9.28 we note the following:

Key point

When calculating partial fractions:

- the denominator must be factorised

- each factor of the denominator generates a partial fraction

- for repeated factors, the partial fractions are generated by the factor and the square of the factor

There are more complicated examples of partial fractions that are not dealt with in this chapter. For example, some fractions have factors in the denominator that are quadratics that will not factorise. For a more in-depth study of partial fractions see *Mathematics for Engineers: A Modern Interactive Approach*, 3rd edition, by A. Croft and R. Davison (2008, Pearson Education).

Exercise 9.5

1. Find the partial fractions of the following:

   (a) $\dfrac{7x + 18}{(x + 2)(x + 3)}$ (b) $\dfrac{2x - 7}{x^2 + 5x + 4}$ (c) $\dfrac{-9}{2x^2 + 15x + 18}$ (d) $\dfrac{5x - 11}{x^2 - 5x + 4}$ (e) $\dfrac{3x + 11}{2x^2 + 3x - 2}$

2. Find the partial fractions of

   $$\dfrac{x + 21}{2(2x + 3)(3x - 2)}$$

3. Find the partial fractions of

   (a) $\dfrac{x - 35}{x^2 - 25}$ (b) $\dfrac{x - 4}{x^2 - 6x + 9}$ (c) $\dfrac{5x + 4}{-x^2 - x + 2}$ (d) $\dfrac{12x - 5}{9x^2 - 6x + 1}$

Test and assignment exercises 9

1. Simplify

   (a) $\dfrac{5a}{4a + 3ab}$ (b) $\dfrac{5a}{30a + 15b}$ (c) $\dfrac{5ab}{30a + 15b}$ (d) $\dfrac{5ab}{ab + 7ab}$ (e) $\dfrac{y}{13y + y^2}$ (f) $\dfrac{13y + y^2}{y}$

2. Simplify the following:

   (a) $\dfrac{5a}{7} \times \dfrac{14b}{2}$ (b) $\dfrac{3}{x} + \dfrac{7}{3x}$ (c) $t - \dfrac{4 - t}{2}$ (d) $4(x + 3) - \dfrac{(4x - 5)}{3}$

   (e) $\dfrac{7}{x} + \dfrac{3}{2x} + \dfrac{5}{3x}$ (f) $x + \dfrac{3x}{y}$ (g) $xy + \dfrac{1}{xy}$ (h) $\dfrac{1}{x + y} + \dfrac{2}{x - y}$

3. Simplify the following:

   (a) $\dfrac{y}{9} + \dfrac{2y}{7}$ (b) $\dfrac{3}{x} - \dfrac{5}{3x} + \dfrac{4}{5x}$ (c) $\dfrac{3x}{2y} + \dfrac{5y}{6x}$ (d) $m + \dfrac{m + n}{2}$

   (e) $m - \dfrac{m + n}{2}$ (f) $m - \dfrac{m - n}{2}$ (g) $\dfrac{3s - 5}{10} - \dfrac{2s - 3}{15}$

4. Simplify

   $$\dfrac{x^2 - x}{x - 1}$$

5. Find the partial fractions of

(a) $\dfrac{4x + 13}{(x+2)(x+7)}$
(b) $\dfrac{x+4}{(x+1)(x+2)}$
(c) $\dfrac{x-14}{(x+4)(x-5)}$
(d) $\dfrac{-x}{(3x+1)(2x+1)}$

(e) $\dfrac{6x-13}{(2x+3)(3x-1)}$

6. Calculate the partial fractions of

(a) $\dfrac{8x+19}{x^2+5x+6}$
(b) $\dfrac{3x+11}{x^2+9x+20}$
(c) $\dfrac{3x+7}{x^2+4x+4}$
(d) $\dfrac{x-6}{x^2-6x+9}$
(e) $\dfrac{8x-7}{4x^2-4x+1}$

Transposing formulae

When presented with a formula, there is usually one expression which is equal to another, for example in the formula for the area of a circle: $A = \pi r^2$, the area is denoted by A, and here it is the subject of the equation. It is equal to the product of π, and the radius squared.

You may need to rearrange, or transpose, the equation so that one of the other parts of the equation becomes the subject. The subject is traditionally written on the left hand side of the equation.

Some examples to demonstrate some basic rules of rearranging equations:

$$y = x + 1$$

If you wanted to rearrange this so that x was the subject you first need to get x on its own. Here 1 is added to x to give a value for y. If you take 1 away from both sides:

$$y - 1 = x + 1 - 1$$

Then, on the right hand side of the equation, +1 and -1 cancel out:

$$y - 1 = x \cancel{+1} \cancel{/1}$$

leaving you with: $y - 1 = x$

By convention the equation is expressed with the subject on the left hand side: $x = y - 1$

The equals sign is like a pivot. It doesn't matter which way round the equation is written as long as what is on opposite sides of the equals sign remains the same.

So $y = mx + c$ is exactly the same as $mx + c = y$.

Take the example $y = mx$. If you wanted to make x the subject here, x is multiplied by m to give a value for y, so if you divide each side by m, it can be cancelled out of its association with x:

$$\frac{y}{m} = \frac{\cancel{m}x}{\cancel{m}}$$

leaving you with $\frac{y}{m} = x$ which needs to be expressed as $x = \frac{y}{m}$.

If the equation has both multiplication and addition elements in it you need to work so that you isolate the part of the equation you are trying to make the subject. So in the equation:

$$y = mx + c$$

if you are trying to make x the subject you first remove the most distant parts of the equation from x, in this case start by removing $+ c$. You would do this by subtracting c from both sides, and this would cancel out on the right hand side of the equation:

$$y - c = mx \cancel{+ c} \cancel{- c}$$

giving you: $y - c = mx$

Now x is just with m on the right hand side, so you can divide both sides of the equation by m, which allows you to cancel m from the right hand side:

$$\frac{y - c}{m} = \frac{\cancel{m}x}{\cancel{m}}$$

leaving: $\frac{y-c}{m} = x$ which you rewrite as $x = \frac{y-c}{m}$.

Going back to the equation for the area of a circle: $A = \pi r^2$, if you wanted to make r the subject of the equation, first you need r on its own, which means removing π from the right hand side of the equation. You do this by dividing both sides by π and cancelling out π from the right hand side.

$$\frac{A}{\pi} = \frac{\cancel{\pi} r^2}{\cancel{\pi}}$$

leaving: $\frac{A}{\pi} = r^2$. Where you have a squared expression in an equation, in this example r on the right hand side is squared, you need to take the square root of r^2. This gives you r, but, as with any equation, what you do to one side, you must do to the other. So if you take the square root of the right hand side you must also take the square root of the left hand side:

$$\sqrt{\frac{A}{\pi}} = \sqrt{r^2}$$

So the square root of $r^2 = r$, giving an answer of:

$$\sqrt{\frac{A}{\pi}} = r \quad \text{or} \quad r = \sqrt{\frac{A}{\pi}}$$

Examples involving brackets:

Rearrange $y = 4(x - 3)$ to make x the subject. There are several ways you could tackle this. Here is one example.

You first need to get x out of the bracket, and that can be done by expanding the bracket

$$y = 4x - 12$$

Then add 12 to both sides so that the +12 and -12 on the right hand side cancel out:

$$y + 12 = 4x - \cancel{12} + \cancel{12}$$

So $y + 12 = 4x$

Then divide both sides by 4 and cancel the 4s on the right hand side:

$$\frac{y + 12}{4} = \frac{\cancel{4}x}{\cancel{4}}$$

This gives: $\frac{y+12}{4} = x$. You could simplify this by separating the fraction on the left hand side:

$\frac{y}{4} + \frac{12}{4} = x$ and then simplify $\frac{12}{4} = 3$ so you could express the answer as:

$$x = \frac{y}{4} + 3$$

Another example:

Rearrange the following equation to make x the subject:

$$y = \frac{4 + t}{3 - x}$$

As before, there are different ways of going about this. Here is one example of how you could approach this problem. The first thing to do is to get the 3 - x on to the top of a fraction. The best thing to do is put brackets round the 3 - x because then it is treated as one expression until it is time to separate it. Also you can put brackets around the 4 + t because it can be treated as one expression throughout. Both sides are then multiplied by (3 - x) and so this expression can be cancelled from the right hand side of the equation.

$$y(3 - x) = \frac{(4 + t)(3 \cancel{- x})}{(3 \cancel{- x})}$$

You can then expand the brackets on the left hand side so you can get at the x.

$$3y - xy = (4 + t)$$

To get the expression with x in on its own, because it is $3y$ minus xy, you can take $3y$ away from both sides:

$$\cancel{3y} - \cancel{3y} - xy = (4 + t) - 3y$$

Cancelling $3y$ from the left hand side of the equation gives you:

$$-xy = (4 + t) - 3y$$

Now you need to divide each side by $-y$ to get x on its own:

$$\frac{\cancel{-}yx}{\cancel{-}y} = \frac{(4 + t) - 3y}{-y}$$

$$x = \frac{(4 + t) - 3y}{-y}$$

Another example: Rearrange the following to make x the subject:

$$y = \frac{3s}{4(x - 7)}$$

As with other examples, there will be several ways of tackling this problem. The solution here is just one possibility. Multiply both sides by $(x - 7)$ to get x to the top of the equation and cancel this expression from the right hand side. This leaves you with:

$$4y(x - 7) = \frac{3s}{4}$$

Divide both sides by y and cancel y from the left hand side of the equation:

$$(x - 7) = \frac{3s}{4y}$$

Remove the brackets from the left hand expression and add 7 to both sides:

$$x - 7 + 7 = \frac{3s}{4y} + 7$$

On the left hand side the $7x$ cancels out, leaving you with:

$$x = \frac{3s}{4y} + 7$$

Examples of rearranging formulae:

$$\text{force} = \text{mass} \times \text{acceleration}$$

If you needed to rearrange to make mass the subject, you would divide both sides by acceleration:

$$\frac{\text{force}}{\text{acceleration}} = \frac{\text{mass} \times \text{acceleration}}{\text{acceleration}}$$

On the right hand side the two accelerations cancel out. You then rewrite it so that mass is on the left side of the equation

$$\text{mass} = \frac{\text{force}}{\text{acceleration}}$$

Take the equation $P = \rho R_d T$, where

ρ = density

R_d = gas constant for dry air

T = temperature

If you were to rearrange the equation to make density the subject you would divide both sides by R_d and T

$$\frac{P}{R_d T} = \frac{\rho R_d T}{R_d T}$$

This allows you to cancel R_d and T from the right hand side of the equation.

$$\frac{P}{R_d T} = \rho$$

In the equation $C = 2\pi r$ you are more often provided with the circumference and you will need to calculate the radius. This is because it is almost impossible to accurately determine the centre of a circle but you can measure the circumference. To rearrange the equation to make the radius the subject you need to divide both sides of the equation by 2π.

$$\frac{C}{2\pi} = \frac{2\pi r}{2\pi}$$

You can then cancel 2π from the left hand side.

$$\frac{C}{2\pi} = r$$

Test your understanding

1. Rearrange the following to make s the subject:

a. $t = sm$

b. $t = sm - p$

c. $t = 2(s + m)$

d. $t = m(s - p)$

2. Rearrange the following to make b the subject

a. $g = ab$

b. $g = ac + b$

c. $g = \dfrac{c+a}{b}$

d. $g = \dfrac{2(b-1)}{c}$

3. Rearrange the following to make m the subject

a. $s = m^2$

b. $s = \sqrt{m + f}$

c. $s = 2f - \sqrt{m}$

d. $s = \dfrac{f+2}{m^2}$

4. Rearrange to make w the subject:

a. $d = \dfrac{(w+2)}{x^3}$

b. $d = w^3 + p$

c. $d = \sqrt[3]{wp}$

d. $d = \dfrac{(wp)^3}{2}$

5. Rearrange to make *mass* the subject of the following:

a. $moles = \dfrac{mass}{molar\ mass}$

b. $molarity = \dfrac{mass}{molar\ mass \times volume}$

6. Rearrange $A = \varepsilon cl$ to make c the subject.

7. Rearrange to make r the subject of the equation: $V = \dfrac{4}{3}\pi r^3$

8. Take the following formula: $v = \dfrac{V_{max}[S]}{K_m+[S]}$ to make the following the subject

a) V_{max}

b) K_m

Measurements

Units

Measurements, when dealing with dimensions, don't mean anything without units. Units are vital in communicating your results and helping you work out if you are calculating something properly.

You need to know these prefixes:

giga-	10^9
mega-	10^6
kilo-	10^3
1	1
milli-	10^{-3}
micro-	10^{-6}
nano-	10^{-9}
pico-	10^{-12}
femto	10^{-15}

SI units

There are seven units, known as SI units, from which all others are derived:

Length	meter	m
Time	second	s
Mass	kilogramme	kg
Electric current	ampere	A
Thermodynamic temperature	kelvin	K
Amount of substance	mole	mol
Luminous intensity	candela	cd

These are the universally accepted units of science.

Other concepts have units with their base in the SI units. Examples are:

Area	m^2
Volume	m^3
Density	$kg\ m^{-3}$
Velocity, speed	$m\ s^{-1}$
Acceleration	$m\ s^{-2}$

Ideally you would work in SI units, and always express numbers in their SI units. However, common usage means that, for example, grams are often used instead of kg, and minutes or hours instead of seconds; this is acceptable. However, it is preferable to avoid quoting answers in units such as cm when presenting data in a results section, on graphs or tables for example. The overriding argument though, is for clear communication of data and this should be the most important factor when deciding how to present your results.

How to avoid lots of zeros when converting to SI units

If you carry out a calculation and the answer is, for example 600 km and you wanted to convert this to the SI units m:

1000 m = 1 km

So you would multiply 600 m by 1000 giving 600 000 m.

Rather that use all these zeros, you can express 600 000 as 6×10^5 m and this makes for clearer communication of your results.

Calculations often require you to convert between units, for example, to have the same units on both sides of an equation. If you needed to express 0.054 kg in g, because there are 1000 g in 1 kg, you would multiply 0.054 by 1000:

0.054 kg = 54 g

You may need to convert between units at the end of a calculation, to provide a clearer communication of your results. To avoid making mistakes, only convert between units at the beginning or end of a calculation, NEVER in the middle

An example: An elephant walked 300 m in 12 h, how many km will it walk in 20 days?

First of all the distance:
The question is in m, the answer is asked for in km. So you need to convert 300 m into km.

There are 1000 m in 1 km, or 0.001 km = 1m.
So 300 m needs to be multiplied by 0.001 to convert it to km, 300 m = 0.3 km

Secondly the time:
You need to convert 12 h into days.

There are 24 hours in one day, so $\frac{12}{24} = 0.5$ days

Then you need to look at the question like a proportion. If the elephant walks 0.3 km in 0.5 days, how many multiples of 0.3 km will it walk in 20 days.

So how many 0.5 days fit in 20 days, or $\dfrac{20\text{ days}}{0.5\text{ days}}$

It is this multiple of 0.3 km, so: $\dfrac{20\text{ days}}{0.5\text{ days}} \times 0.3\text{ km} = 12\text{ km}$

The answer is: the elephant walks 12 km in 20 days.

Another example: A fungal hypha measured 0.043 mm and two hours later it measured 0.862 mm. What is the rate of growth of the hypha in μm min⁻¹.

Growth in 2 h $= 0.862 - 0.043 = 0.819$ mm

In 2 h it grew 0.819 mm. The question is essentially asking what fraction of two hours is one minute.

2 h x (60 x 2) = 120 min

So 2 h is equivalent to 120 min.

$$\dfrac{1\text{ min}}{120\text{ min}} \times 0.819\text{ mm} = 0.006825\text{ mm}$$

Then you need to convert mm into μm. There are 1000 μm in 1 mm, so you need to multiply 0.006825 mm by 1000.

0.006825 mm min⁻¹ x 1000 = 6.825 μm min⁻¹

In both these examples you are looking at proportions:

In the elephant example you could express the situation as $\dfrac{300\text{ m}}{12\text{ h}} = \dfrac{12\text{ km}}{20\text{ days}}$

For the fungal hyphae you could write it as $\dfrac{0.819\text{ mm}}{2\text{ h}} = \dfrac{6.825\text{ μm}}{1\text{ min}}$

To work out any calculation like this you need to work out which bit of this equation it is that you need to find out, and convert units as appropriate so that they match each side. You need to be able to convert between units with ease, and should practise this skill if you find it difficult.

Working in areas

Rectangles

You may need to calculate the area of a leaf which you are using to measure photosynthesis, or the area of tissue you have stained, or the area over which you are counting bacteria colonies. Most biological shapes are irregular, but it is acceptable to make an approximation to standard shapes. For example, the area of a palisade cell in a leaf might approximate to a rectangle. You could measure the length, l, and width, w and find the area.

For example: A palisade cell had a length of 8.2 µm and a width of 3.4 µm. Calculate its surface area.

$$A = l \times w$$
$$A = 8.2 \times 3.4$$

Area of the cell = 27.88 µm²

Triangles

Some shapes can approximate to triangles. Here you need the length, l, and the height, h, or width, w, depending on the object.

For example: Calculate the surface area of a leaf which has a length of 3.6 cm and a maximum width of 2.9 cm

Figure 11.1
Approximating the area
of a leaf to a triangle

this leaf can approximate to a triangle.

$$A = \frac{1}{2}bh$$

$$A = \frac{1}{2} \times 3.6 \times 2.9 = 5.22 \text{ cm}^2$$

The trapezium

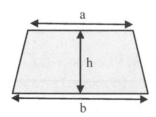

Figure 11.2
A trapezium

$$\text{Area} = \frac{1}{2}(a + b)h$$

Circles

The area of a circle can be used in biological situations for example, the field of view under a microscope or the surface of a petri dish:

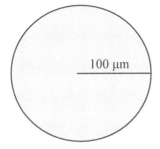

$100\ \mu m$

Area $= \pi r^2$

Area $= \pi \times 100\ \mu m \times 100\ \mu m$

Area $= 31420\ \mu m^2$

However, with circles it is rarely the radius that is measured. This is because it is almost impossible to estimate the centre of a circle accurately. You will usually be measuring the circumference or the diameter of the circle.

If the diameter is measured, the radius is simply half the diameter.

If the circumference is measured the radius can be determined by rearranging the equation where C is the circumference:

$$r = \frac{C}{2\pi}$$

For example:

What is the cross sectional area of an ice core which has a circumference of 10.7 cm?

$$r = \frac{C}{2\pi} = \frac{10.7}{2\pi} = 1.702958\ cm$$

$$A = \pi r^2 = \pi(1.702958)^2 = 9.11\ cm^2$$

Note that you never round up in the middle of an equation. When you have finished your calculation look back at your data and think about how accurate you can claim to be in your answer. Here the data you were given was given to one tenth of a cm. The answer is given to one hundredth of a cm², which is a reasonable claim of accuracy.

Sector of a circle

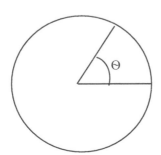

Θ

Area $= \dfrac{1}{2}\pi r^2\theta$

For most though, the problem comes when converting between units in areas. Mistakes are often made through guessing, trying to convert measurements without properly working it out. Avoid short cuts and you shouldn't go wrong. For example, if you are faced with any conversion that involves an area, draw yourself a square and work out the conversion.

For example, convert 0.004 km^2 to m^2

First work out how many m^2 there are in 1 km^2:

.

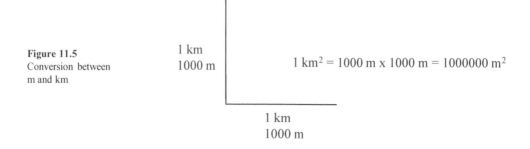

Figure 11.5
Conversion between
m and km

1 km
1000 m

1 km^2 = 1000 m x 1000 m = 1000000 m^2

1 km
1000 m

So you need to multiply 0.004 km^2 by 1000 000 = 4000 m^2 or   4 x 10^3 m^2

Note that as the units get smaller, i.e. metres are smaller than kilometres, so the number gets bigger.

Another example: Convert 6 x 10^5 μm^2 to mm^2.

Figure 11.6
Conversion between
mm and μm

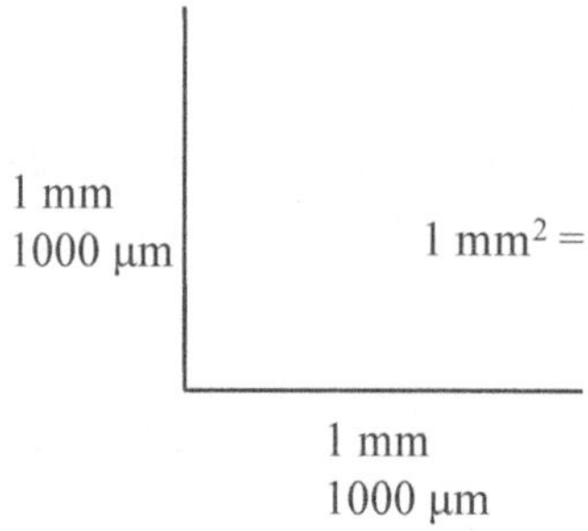

1 mm
1000 μm

1 mm^2 = 1000 μm x 1000 μm = 1000000 mm^2 = 10^6 mm^2

1 mm
1000 μm

So you multiply 6 x 10^5 μm^2 by 10^{-6} (or divide it by 10^6) to get 6 x 10^{-1} mm^2 or 0.6 mm^2

Note that here the units are getting larger, i.e. millimetres are larger than micrometres, so the number gets smaller.

Example: If you found two beetles in an area 4 x 10^4 mm², assuming an even distribution, how many would you expect to find in 1 m²?

Firstly, 4 x 10^4 mm² is 40 000 mm², or 200 mm x 200 mm.

The question is asking how many of these 200 x 200 mm² would you get in 1 m². When in doubt, draw yourself a diagram.

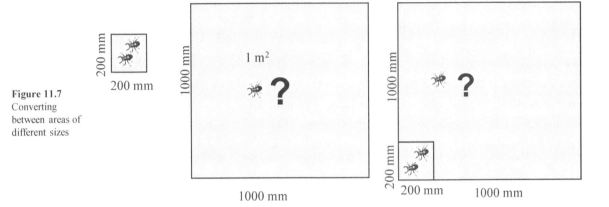

Figure 11.7
Converting
between areas of
different sizes

1 m² = 1000 000 mm² or 1 x 10^6 mm

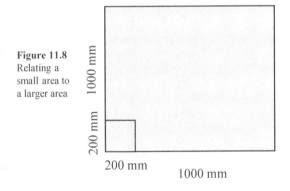

Figure 11.8
Relating a
small area to
a larger area

The big area is $\frac{1000\,mm}{200\,mm} \times \frac{1000\,mm}{200\,mm}$ bigger than the smaller area.

$$= \frac{1000000\ mm^2}{40000\ mm^2} = 25 \text{ times bigger}$$

If the area is 25 times bigger, there must be 25x more beetles in the larger area compared to the smaller area, so multiply the number of beetles by 25:

2 x 25 = 50 beetles.

Answer: You would expect to find 50 beetles in 1 m².

The general equation is:

$$\frac{\text{large area}}{\text{number in the large area}} = \frac{\text{small area}}{\text{number in the small area}}$$

That is, that the number that are found in the small area is exactly proportional to the number that are in the large area – they are at the same concentration, essentially, assuming that the distribution is even. You can then rearrange this to work out which part of the equation you don't know.

You can rewrite this as the equation:

$$\frac{\text{large area}}{\text{small area}} \times \text{number in the small area} = \text{number in large area}$$

Plugging in the numbers from the previous example (being careful to get the units right):

$$\frac{1000000 \text{ mm}^2}{40000 \text{ mm}^2} \times 2 \text{ beetles} = 50 \text{ beetles}$$

Another example: If there were 5 fungal spore patches on 16 mm² of bark, assuming an even distribution on the tree, how many spore patches would there be on 2 m² of bark?

$$\frac{\text{large area}}{\text{small area}} \times \text{number in the small area} = \text{number in the large area}$$

The large area is 2 m². You need to convert this to mm². There are 1000 000 mm² in 1 m². So 2 m² is equivalent to 2 000 000 mm².

$$\frac{2000000 \text{ mm}^2}{16 \text{ mm}^2} \times 5 \text{ fungal patches} = 625000 \text{ fungal patches}$$

The answer would best be displayed as 6.25×10^5 fungal patches on 2 m² of bark.

Another example: If you knew there were 700 000 corals on a coral reef measuring 2.6 km², how many might you expect to find if only observing 30 m²?

$$\frac{\text{small area}}{\text{large area}} \times \text{number in the large area} = \text{number in the small area}$$

There are 10^6 m² in 1 km², so 2.6 km² is equivalent to 2.6×10^6 m².

$$\frac{30 \text{ m}^2}{2.6 \times 10^6 \text{m}^2} \times 700000 \text{ corals} = 8 \text{ corals}$$

So you might expect to find 8 corals in 30 m².

Working in volumes

You may need to calculate the volume of a cell so that you can determine whether cell volume changes with a change in experimental conditions, or you may need to calculate the volume of a tree trunk to determine increase in biomass over time.

As with areas, you can make approximations to geometric shapes. For example many protist cells approximate to a sphere and many plant cells approximate to a cuboid.

Cuboids

For a cuboid the volume is found by multiplying the height, width and depth, for example:

Figure 11.9
Diagrammatic
representation of
a plant cell

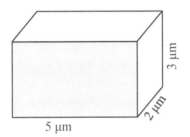

Volume of the cell = $5 \times 2 \times 3 = 30 \ \mu m^3$

Another example: Find the volume of a fish tank measuring 3.0 m x 1.4 m x 2.2 m.

Volume = 3.0 x 1.4 x 2.2 = 9.24 m³

Cylinders

Some shapes can approximate to a cylinder. Here the volume can be worked out by multiplying the surface area, which is a circle, by the length, or height.

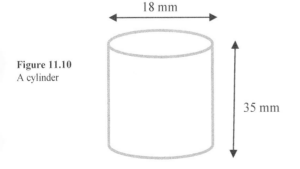

Figure 11.10
A cylinder

A nematode worm could be approximated to a cylinder. For example, calculate the volume of a nematode worm which is 35 mm long and has a diameter of 18 mm.

$V = \pi r^2 h$

$V = \pi 9^2 \times 35$

$V = 8906.42 \ mm^2$

Volume $= 8906 \ mm^2$

Spheres

Many objects can approximate to a sphere, for example, planets, puff balls, certain protist cells. You may, for example, need to determine the volume of a protist cell which can approximate to a sphere, with a diameter of 6 μm:

$$\text{Volume of a sphere} = \frac{4}{3}\pi r^3$$

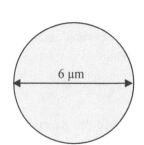

Figure 11.11
An approximate shape for a spherical protist cell

6 μm

The radius is half the diameter:

$$\frac{6\,\mu m}{2} = 3\,\mu m$$

So $\text{Volume} = \frac{4\times\pi\times3\mu m\times3\mu m\times3\mu m}{3} = 339.3\ \mu m^3$

Note – the equation above is a way of putting the numbers into your calculator so that you don't make a mistake with what is raised to the power 3.

Another example: Find the volume of a puff ball, which can approximate to a sphere, with a circumference of 24.3 cm.

$$C = 2\pi r \qquad r = \frac{C}{2\pi} \qquad r = \frac{24.3}{2\pi} \qquad r = 3.867465\ \text{cm}$$

$$\text{Volume of sphere} = \frac{4}{3}\pi r^3$$

$$V = \frac{4}{3}\pi(3.867465)^3 = 242.3\ \text{cm}^3$$

Cones

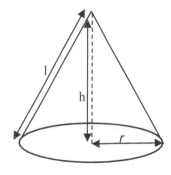

Figure 11.12
A cone

l

h

r

$$\text{Volume} = \frac{1}{3}\pi r^2 h$$

$$\text{Curved surface area} = \pi r l$$

Converting between units when working in volumes

You may need to convert between volumes. If, for example, the previous question asked you to give your answer in m^3, you would need to convert between cm^3 and m^3.

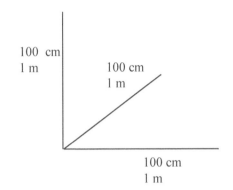

Figure 11.13
Converting
between cm
and m

1 m³ = 100 x 100 x 100 cm³
1 m³ = 100000 cm³

The units are getting bigger, so the number will get smaller. You need to multiply your answer by 10^{-6}, so the answer is 2.423 x 10^{-4} m³

More examples:

Convert 3.5 x 10^7 m³ to km³.
There are 1000 000 000 m³ in 1 km³, or 10^9 m³ in 1 km³. So you need to divide 3.5 x 10^7 by 10^9, or multiply by 10^{-9}. The units are getting larger, so the number gets smaller.
Answer: 3.5 x 10^{-2} km³

Convert 6.23 x 10^{-12} mm³ to μm³.
There are 1000 000 000 μm³ in 1 mm³, or 10^9 μm³ in 1 mm³. So you need to multiply 6.23 x 10^{-12} by 10^9. The units are getting smaller, so the number gets bigger.
Answer: 6.23 x 10^{-3} μm³

Convert 3.27 x 10^{-8} cm³ to μm³.
There are 10 mm in a cm and 1000 μm in a mm, so in each dimension you need to take into account that there are 10,000 μm in 1 cm. So in 1 cm³ there are 10,000 x 10,000 x 10,000 μm³.
So you need to multiply 3.27 x 10^{-8} by 10^{12}. The units are getting smaller, so the number gets bigger.
Answer: 3.27 x 10^4 μm³

Convert 6.1 x 10^7 mm³ to dm³.
There are 100 mm in a dm so in 1 dm³ there are 100 x 100 x 100 mm³.
So you need to multiply 6.1 x 10^7 by 10^6. The units are getting larger, so the number gets smaller.
Answer: 6.1 x 10^1 dm³ or 61 dm³

An example of a situation where this sort of conversion is needed:

If the volume of a plant cell is 8000 μm³, how many cells would be present in a volume of leaf 1 mm³?

First convert the units, converting mm³ to μm³. If you are not confident about conversions, draw yourself a diagram:

Figure 11.14
Converting between
mm and μm

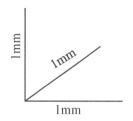
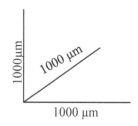

1mm³=1000mm x 1000 mm x 1000 mm

1mm³ = 1000 000 000 μm³

Or 1 x 10⁹ μm³

The problem is one of proportions, being the proportion of the big volume over the smaller volume:

$$\frac{\text{large volume}}{\text{number in the large volume}} = \frac{\text{small volume}}{\text{number in the small volume}}$$

You then sort out which is the bit of the equation you need to find out and rearrange the equation to make this the subject.

$$\frac{\text{big volume}}{\text{little volume}} \times \text{number of plant cells in the little volume} = \text{number of plant cells in the big volume}$$

$$\frac{1000000000 \ \mu m^3}{8000 \ \mu m^3} \times 1 \text{ plant cell} = 125000 \text{ cells}$$

With all problems such as this, it is really important to look at your answer and look back at the question and check the answer makes sense. Take the above equation and cancel by factors of 10 and estimate your answer. Then when you use a calculator to work out the answer you will know if you are in the right ball park.

$$\frac{1000000\cancel{000} \ \mu m^3}{8\cancel{000} \ \mu m^3} \times 1 \text{ plant cell}$$

Volume conversions can apply to liquids too. For example: A biologist has put 0.1 g salt into 125 ml water. What is the concentration of salt in g l^{-1}?

The small volume is 125 ml, the big volume is 1 litre. The first thing is to convert units, converting ml into litres. If you are unsure, draw a diagram:

Figure 11.15
A 10 cm cube

1 ml = 1 cm^3

1 litre is a volume 10 cm x 10 cm x 10 cm = 1000 cm^3

So 1 litre = 1000 ml

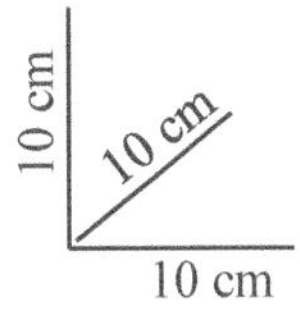

$$\frac{\text{big volume}}{\text{little volume}} \times \text{mass in the little volume} = \text{mass in the big volume}$$

$$\frac{1000 \text{ ml}}{125 \text{ ml}} \times 0.1 \text{ g} = 0.8 \text{ g}$$

The answer is: the concentration of salt would be 0.8 g in 1 litre, or 0.8 g l^{-1}.

Another example: The mass of copper sulphate in a small pond was 3.6 g in 40 000 litres. To mimic this in a flask in the laboratory, what mass of copper sulphate would be needed in a volume of 200 ml?

This is a question about proportions:

$$\frac{\text{pond volume}}{\text{mass in the pond}} = \frac{\text{flask volume}}{\text{mass in the flask}}$$

Work out what you need to find out, and rearrange the equation. Then plug in the numbers, making sure the units are equivalent, so 200 ml needs to be converted to 0.2 litres.

$$\frac{\text{flask volume}}{\text{pond volume}} \times \text{mass in the pond} = \text{mass in the flask}$$

$$\frac{0.2 \text{ litres}}{40\ 000 \text{ litres}} \times 3.6 \text{ g} = 0.000018 \text{ g}$$

A better way to express this answer would either be in standard form: 1.8×10^{-5} g
or by converting to μg. There are 10^6 μg in 1 g, so the answer needs to be multiplied by 10^6, giving 18 μg.

The previous question might have been phrased in a slightly different way:

The concentration of copper sulphate in a small pond was 90 µg l⁻¹. How much copper sulphate would be needed to be added to 200 ml water to achieve the same concentration?

Concentration here is mass of copper in 1 litre of the pond water. Both sides need to be in the same units

$$\frac{\text{mass in the pond}}{\text{pond volume}} = \frac{\text{mass in the flask}}{\text{flask volume}}$$

$$\frac{90 \ \mu g}{1 \ \text{litre}} = \frac{\text{mass in the flask}}{0.2 \ \text{litres}}$$

$$\frac{90 \ \mu g}{1 \ \text{litre}} \times 0.2 \ \text{litres} = \text{mass in the flask}$$

The answer is: You would need 18 µg copper sulphate to achieve the same concentration.

In some questions you might need to work in volumes such as cm^3 or m^3, and also litres.

For example:
A cuboid fish tank measured 3.5 m x 4.6 m x 2.8 m. How much water, in litres, would be needed to fill the tank completely?

The volume of the fish tank would be: $V = 3.5 \times 4.6 \times 2.8 = 45.08 \ m^3$

1 litre $= 1 \ dm^3$ so you need to convert 45.08 m^3 to dm^3. There are 10 dm in 1 m, so 1000 dm^3 in 1m^3. The units are getting smaller so the number will get bigger. You need to multiply 45.08 m^3 by 1000.

The volume of the tank is $45.08 \times 1000 = 4.508 \times 10^4 dm^3$
So the amount of water needed would be 4.508×10^4 litres

Note that: 1 litre $= 1 \ dm^3$
                                              1 ml $= 1 \ cm^3$

Deciding how many decimal places and significant figures to use

When you carry out a calculation using a calculator it might give you an answer to many significant figures. While carrying out a calculation it is a good idea to use many significant figures. However, it is important to give the final answer with the number of significant figures that reflects the accuracy of the initial data.

For example:

You need to work out the volume of a spherical cell. You measured the diameter and found it to be 6.3 μm. This means that you know its diameter to an accuracy of 0.1 μm; you know it is not 6.2 μm, nor 6.4 μm. It could be 6.34 μm, but the accuracy of your measuring does not provide you with a figure accurate to 0.01μm.

The radius is half the diameter, so is 3.15 μm. It is okay to use more significant figures at this stage because it is the middle of the calculation.

Volume of a sphere is $V = \frac{4}{3}\pi r^3$ so $V = \frac{4}{3} \times \pi \times 3.15 \times 3.15 \times 3.15$

$V = 130.924303 \ \mu m^3$

You should not give this as your final answer. It shows that you have not thought about your answer and what it means. You have just copied the numbers from your calculator. It is all about communicating your answer to others, and your answer is more than just the numbers. The answer should also reflect your confidence in the accuracy of the number, and you can do this through the number of significant figures you show.

Your original data was accurate to 0.1 μm. Your answer must reflect this, and so be given to one decimal place:

$$V = 130.9 \ \mu m^3$$

In this way you are communicating that you know the volume of the cell is not 131.0 μm nor 130.8 μm, and this is as accurate as you can be, given the initial measurement. The answer from the calculator, to five decimal places, gives the impression you can be accurate to ten pico meters (pm), and this is clearly not true.

What happens if you round up a number in the middle of a calculation?

Taking the same example, if you calculated the radius of the cell as half the diameter, and rounded 3.15 μm to 3.2 μm before calculating the volume, your calculation would be:

$$V = \frac{4}{3} \times \pi \times 3.2 \times 3.2 \times 3.2$$
$$V = 137.25827743$$

$$V = 137.3 \ \mu m^3$$

Here you can see the importance of only rounding up or down at the end of calculation, because the answer is very different if you round up or down in the middle of the calculation. This second answer would be wrong.

Test your understanding.

Carry out the following conversions:

1. Convert to mm:

a. 1200 μm

b. 0.4 cm

c. 0.0035 m

2. Convert to g:

a. 0.045 kg

b. 36 mg

c. 64000μg

3. Convert to m²:

a. 3500 mm²

b. 480000 μm²

c. 0.0005 km²

4. Convert to mm³

a. 0.000004 m³

b. 853000 μm³

c. 0.004 cm³

5. Convert to ml:

a. 34 cm³

b. 5 l

c. 680 μl

6. Convert to litres:

a. 43 dm³

b. 3×10^2 cm³

c. 4.6 dm³

7. Convert to μm²:

a. 9.3×10^5 nm²

b. 7.47×10^{-8} mm²

c. 4.71×10^{-6} cm²

8. Convert to km²:

a. 3.51×10^{20} μm²

b. 8.96×10^8 m²

c. 7.64×10^{11} mm²

9. Convert to mm³

a. 7.92×10^9 μm³

b. 8.335×10^{15} nm³

c. 2×10^{-5} cm³

10. Convert to km³:

a. 4.3×10^7 dm³

b. 9.46×10^{13} cm³

c. 3.7×10^{20} μm³

11. Calculate:

a. The area of a square with sides of 4.2 mm.

b. The area of a quadrat with sides of 0.56 m.

c. The volume of a cube with sides of 4.3 cm.

d. The area of a circle with a radius of 7 mm.

e. The area of a circle with a diameter of 2.9 m.

f. The area of a circle with a circumference of 1.6 μm.

g. The volume of a sphere with a radius of 5.75 m.

h. The volume of a sphere with a circumference of 0.23 m.

i. The volume of a sphere with a diameter of 5 km.

j. The volume of a cylinder with a radius of 7 μm and a length of 10 μm.

k. The volume of a cylinder with a circumference of 3.7 m and a height of 6.6 m.

l. The volume of a cone with a height of 4.1 cm and a radius of 2.9 cm.

12. Calculate the following:

a. If a squirrel gathered 18 acorns in 30 min, now many could it gather in 3 hours?

b. If a camel walked 11.6 km across a desert in 10 h, how far would it walk in 3 days?

c. If you had 0.5 g of KCl in 40 ml, what would be the concentration of KCl in g l^{-1}?

d. There is 1 acacia tree in every 500 m^2 of grassland. How many might there be in 4 km^2?

e. If there were 3 cod fish in 1000 litres of sea how many would there be in 4 km^3?

f. If there were 4×10^3 woodlice in $10m^2$ of leaf litter how many would there be in 1 km^2?

g. If a bacterial cell had a volume of 5 μm^3, how many bacteria could you expect to find in a densely packed circular colony on an agar plate? The colony measured 3.5 cm across and had a depth of 1.0 mm.

h. If there were 65 grass plants in a 0.4 m^2 quadrat, assuming even distribution, how many would there be in a field 0.6 km^2?

i. If the volume of a plant cell is 5.9 μm^3, how many would there be in a volume of leaf 4 mm^3?

j. Calculate the maximum volume of water in litres, that a fish tank could hold, that has the dimensions: base 0.4 x 0.2 m and height 0.3 m.

k. A tapeworm releases eggs at a rate of 30,000 per day. Assuming the rate is steady across the 24 h day, how many are released each second?

l. On a woodland floor, evenly and densely packed with bluebells, it was found that there were 280 plants in 1 m^2. Assuming an even distribution, how many would there be in 0.3 km^2?

m. In 0.3 cm^3 of soil there were found to be 3.6×10^2 *Amoeba*. How many would you expect to find in 0.1 m^3?

n. In a field, 0.5 km^2, there was found to be 2×10^6 clover plants. A teacher wanted to be sure that their students found an average of 25 clover plants in each area they surveyed. How big an area would each student be given to study, giving your answer in m^2?

o. How many litres of water would be needed to fill a swimming pool completely, which had a volume of 23 m^3?

13. Calculate the following:

a. A spherical cell has a diameter of 12 μm, what is the volume of this cell?

b. A caterpillar has a circumference of 16 mm and a length of 24 mm. It's shape approximates to a cylinder. What is the volume of the caterpillar?

c. A leaf, approximating to a triangle, has a length of 38 mm and a base width of 29 mm. What is its area?

d. Calculate the biomass of a boletus mushroom, which has a top part that approximates to a half sphere. The mushroom tissues weigh 0.15 g cm^{-3}. The mushroom stalk is 4.0 cm high, and has a circumference of 4.5 cm. The top of the mushroom has a diameter of 6.8 cm.

e. Calculate the mass in mg of a spherical protist cell which has a diameter of 32 μm . The cell has a mass per unit volume of 3 ng μm^{-3}. Calculate the mass of the cell.

f. Calculate the volume of a bacterial colony that contains 5×10^{11} bacteria, where each bacterial cell approximates to a cylinder and has a length of 0.84 μm and a diameter of 0.33 μm.

g. Calculate the mass of a very long flatworm whose shape can approximate to a cuboid, that has a length of 0.89 m, a thickness of 1mm and a width of 5 mm. It has a mass per unit volume of 0.4 g cm^{-3}.

h. Calculate the total volume, in m^3, of earthworms in a field 0.4 km^2. Each worm approximates to a cylinder 7 cm long and 8 mm in circumference and there are 8 worms per m^2.

14. Work out the following. In giving your answer, think about the decimal places and significant figures you are providing.

a. Calculate the speed that a bee flies after you have measured it covering 1.2 m in 3 s.

b. Calculate the area of a rectangular experimental plot where you have measured the length as 4.60 m and the width as 2.34 m.

c. Calculate the volume of an apple, assuming it to be spherical. You have measured the circumference to be 10.2 cm.

d. You have counted 45 cells in an area which you have measured as 12 μm^2. How many would you expect to find in an equivalent area of 250 μm^2.

e. You have measured the diameter of a cylinder to be 8.46 cm and the height to be 5.36 cm. Calculate the area, circumference and volume of the cylinder.

f. Calculate your running speed having run around a circular track with a diameter of 64 m, in 31 s.

$$speed = \frac{distance}{time}$$

g. If you planted a tree and measured its height to be 4.0 m, and monitored the growth of a tree and found that it grew at a steady rate of 1.3 m year^{-1}, how tall would it have been 5 months after you had planted it? How tall would it be after 9 years?

h. A patient came to see you with a rash that covered 12 cm^2 of their back. After a further three days they came back and you found it to be 1.7 times bigger. Assuming it was growing at a steady rate, how many cm^2 was it increasing by each day?

i. You measured the height of a patient and found it to be 1.8 m. You measured their mass and found it was 74 kg. Calculate their body mass index (bmi). The equation for this is: $bmi = \frac{weight\ (kg)}{(height(m))^2}$

j. You measured a small cuboid fish tank to be 20.4 cm x 15.7 cm x 12.8 cm. What volume of water would completely fill it?

k. By using a piece of string, you measured around the outer rim of a trampoline and found its circumference to be 4.75 m. What would the radius of the trampoline be?

l. You measured counted the number of strides while jogging 100 m and found that you had an average stride length of 0.7 m. How many strides would you estimate a pedometer to show after jogging 2.300 km?

m. Using a haemocytometer, you counted the number of cells in a grid three times. The grid measured 0.1 x 1 x 1 mm. The number of cells you counted was 8, 3 and 5. What would your average concentration of cells be per ml?

Dilution and Scaling

Often in biology you need to dilute a solution. It may be a dilution series of a bacteria culture, for example, or a dilution of prey for a feeding experiment. It could be dilution of a nutrient or toxic chemical. It is important to know how to make a dilution series and to calculate the concentration of the substance you are diluting, or the concentrations of the dilutions you are creating. For calculating concentrations of solutions you dilute, see the chapter 'Moles and Molarity'.

For example:

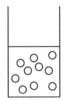

Figure 12.1
Representation of a
bacteria culture.

This is 10 ml of culture where each circle represents 10^5 bacteria. So there are 10×10^5 bacteria in this 10 ml culture, and so the bacteria are at a concentration of 10^5 bacteria ml^{-1}.
You are asked to make a 1/5 (a 1 in 5) dilution.

You could take 1 ml of a 10 ml culture..... and add it to 4 ml of medium

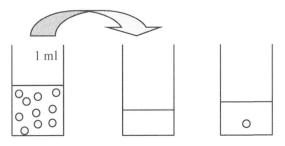

Figure 12.2
Diluting 1 ml of the
bacteria culture

1 ml

You will end up with 5 ml of culture, 1/5th of which was the original culture. There will be 10^5 bacteria in 5 ml, and this will be a concentration of 2×10^4 bacteria ml^{-1}.

You could take 2 ml of the 10 ml culture..... and add it to 8 ml of medium

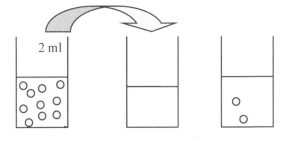

Figure 12.3
Diluting 2 ml of the
bacteria culture

2 ml

You will end up with 10 ml of culture, 1/5th of which was the original culture. You will have twice as much as the previous dilution, but the concentration is the same for both.
i.e. there will be 2×10^5 bacteria in 10 ml, which is a concentration of 2×10^4 bacteria ml^{-1}.

The important values, when working with dilutions, are the properties of the original solution and the amount by which you are diluting that property.

A common exercise is to create a dilution series. This is often where a solution is continually diluted by a factor of 10 each time. Using 10 ml of a bacteria culture at a concentration of 10^8 cells ml^{-1} as an example, 1 ml is taken from the original culture and added to 9 ml of medium. This creates 10 ml of new culture, diluted 1/10 from the original culture, a 10^{-1} (or 1 in 10) dilution. From this new culture, 1 ml is taken and added to another 9 ml of new medium, and so on........

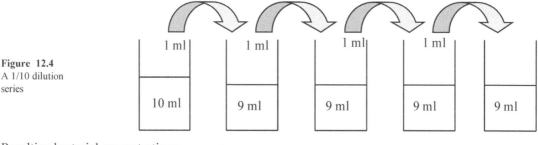

Figure 12.4
A 1/10 dilution series

Resulting bacterial concentrations:

	10^8 cells ml^{-1}	10^7 cells ml^{-1}	10^6 cells ml^{-1}	10^5 cells ml^{-1}	10^4 cells ml^{-1}
Dilution:	1	10^{-1} $(\frac{1}{10})$	10^{-2} $(\frac{1}{100})$	10^{-3} $(\frac{1}{1000})$	10^{-4} $(\frac{1}{10000})$

Each new bacterial solution is 1/10th the concentration of the previous one, because each is a 1/10 dilution.

You may, alternatively, be asked to make a '1 in 2' dilution. This means that one part of the original is mixed with the same volume of new media to end up with a solution twice the volume of the first:

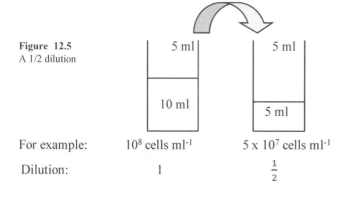

Figure 12.5
A 1/2 dilution

So, for example, the first culture will end up with 5 ml at the original concentration and the second solution will end up with 10 ml at half the concentration of the first.

For example:	10^8 cells ml^{-1}	5×10^7 cells ml^{-1}
Dilution:	1	$\frac{1}{2}$

Colony Forming Units per ml (cfu ml^{-1})

This is a way of counting numbers of bacteria in samples. For example, the number of coliform bacteria in drinking water, or the number of bacteria in a soil sample.

The sample is diluted, using a 10 fold dilution series, and a known volume of each dilution is plated onto suitable media for bacterial cultivation. After incubation, bacterial colonies are counted.

Each bacteria colony derives from one single bacterial cell. So by counting the number of colonies and taking into account the dilution factor and volume plated, the original number of bacteria in the sample can be determined.

Worked examples

1. A protein solution of unknown concentration was diluted 1/5 (a five fold dilution). A protein assay was performed on the diluted protein sample and it was found to have a concentration of 2 mg ml^{-1}. What was the concentration of the original protein solution?

If the diluted sample was 1/5th of the concentration of the original solution (a 1/5 dilution), then you just multiply the concentration by five to find the original concentration:

Answer = 2 mg ml^{-1} x 5 = 10 mg ml^{-1}

2. A solution of bromophenol blue had an absorbance reading at 600 nm of 1. Absorbance is proportional to concentration. This solution was diluted 1/10 (a ten fold dilution). What do you predict the absorbance to be of the diluted bromophenol blue solution?

If you dilute a solution 1/10, its properties such as absorption, will also decrease by 1/10th

So absorption of 1 ÷ 10 = 0.1

3. A 1 ml sample of water from a lake was diluted from 10^{-1} to 10^{-6} and a 100 μl aliquot of the 10^{-6} dilution was plated on bacterial growth media. The bacteria were cultured overnight, and the number of colonies counted. There were 50 colonies counted on the plate.

10^{-6} dilution is a million fold dilution, so firstly multiply 50 by 10^{6}.

$$50 \times 10^6 = 5 \times 10^7 \text{ cfu (100ml)}^{-1}$$

Then you need to take into account that only 100 μl was plated (0.1 ml). This is a tenth of a ml. So you need to multiply by $\dfrac{1}{\text{volume plated (ml)}}$

$$5 \times 10^7 \times \frac{1}{0.1 \text{ ml}} = 5 \times 10^8 \text{ cfu ml}^{-1}$$

4. You are a soil microbiologist and you want to work out how many culturable bacteria there are in 1 g of soil. The 1 g of soil was placed into 10 ml of sterile water. A serial dilution series was performed from 10^{-1} to 10^{-6}. A sample of 100 μl from each dilution was plated onto culture medium and incubated at 30 °C. After three days the colonies growing on the plates were counted and 50 colonies were counted at the 10^{-4} dilution. Using this information, calculate the cfu g^{-1} of soil.

There were 50 colonies at 10^{-4} dilution. This is a 1000 fold dilution.

50 x 1000 = 50,000 cfu (100 μl)$^{-1}$

Then adjust for the volume to get it per ml, i.e. 100 μl x 10 = 1 ml

50,000 x 10 = 500,000 cfu ml^{-1}

1 g of soil was placed in 10 ml and you have worked out how much is in 1 ml. Therefore the total culturable bacteria in 1 g of soil is:

500,000 cfu ml^{-1} x 10 = 5,000,000 cfu g^{-1} (5 x 10^6 cfu g^{-1})

5. A rotifer is feeding on algal cells. You have measured the uptake of algal cells by each rotifer over time with prey at different dilutions of 1/1, 1/2, 1/4, 1/6 prey cells ml^{-1} where the 1/1 dilution was 6000 cells ml^{-1}. These prey concentrations gave you feeding rates of 36, 25, 9 and 6 cells predator^{-1} min^{-1} respectively. How would you represent this on a graph? And how would you express these dilutions as percentages of the original prey concentration?

NB. You would NOT put the dilution series on the x-axis. You would put the actual concentrations of prey on the axis. The dilutions would be 6000, 3000, 1500, 1000 cells ml^{-1} respectively. These concentrations of prey would then be used on a graph:

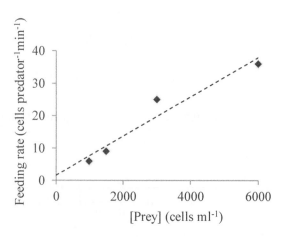

Figure 12.6
Feeding rate against prey concentration

Dilutions are 1/1, 1/2, 1/4 and 1/6. This question simply expects you to convert these fractions into percentages:

1/1 = 100 %   1/2 = 50 %   1/4 = 25 %   1/6 = 16.7%

So, for example, the ¼ dilution gives you a prey concentration that is 25 % of the concentration of the original culture.

Scaling up and down

Often in biology you need to represent one measurement or size by another. For example, a cell is very small and if you make a drawing of it you represent its size using a scale bar. You might wish to represent the length of a gene in a diagram on a sheet of paper where you would need to give an idea of its relative size. You might have a map which represents many kilometres on a page which is less than 10 cm across.

Relative scales can be represented as ratios. You could be told that a diagram is drawn at a scale of 10,000:1, where 1 µm, in reality, is represented on the page by 10,000 µm, or 10 mm. This means that 10 mm on the page is equivalent to 1 µm in real life.

$$\frac{\text{size on the page}}{\text{size in real life}} = \frac{10 \text{ mm}}{1 \text{ µm}}$$

You could be asked to measure something on a diagram and say what its 'real life' measurement is.

For example, if a picture of a cell was drawn to a scale of 10,000:1, then 10 mm on the drawing would be 0.001 mm on the cell.

$$\frac{\text{size of the cell in the picture}}{\text{size of the cell in real life}} = \frac{10 \text{ mm}}{0.001 \text{ mm}}$$

If the mitochondria measured 0.8 mm on the picture, what is its actual size?

The scale is 10,000:1, you know the size on the picture, you need to rearrange the equation so that the object in real life is the subject:

$$\frac{\text{mitochondria in the picture}}{\text{mitochondria in real life}} = \frac{10,000}{1}$$

$$\text{mitochondria in real life} = \frac{1}{10,000} \times \text{mitochondria in real life}$$

$$\text{mitochondria in real life} = \frac{1}{10,000} \times 0.8 \text{ mm} = 0.00008 \text{ mm} \quad \text{or } 80 \text{ nm}.$$

So the mitochondria which appear as 0.8 mm on the picture have an actual size of 80 nm, and you know this because you have been told the picture is at a scale of 10,000:1.

Take a map with a scale of 1:10,000, if a field measured 5 mm across on the map, what would its actual size be. This again is all about ratios.

$$\frac{\text{size on the map}}{\text{size in real life}} = \frac{1}{10,000}$$

Rearrange the equation to make what you don't know as the subject:

$$\text{size in real life} = \frac{10,000}{1} \times \text{size on the map}$$

$$\text{size in real life} = \frac{10,000}{1} \times 5\,\text{mm} = 50000\,\text{mm}$$

Then you need to convert 5000 mm to something more sensible. 50000 mm = 50 m

Graticules

When working with microscopes you may be asked to measure an object under the microscope. To do this you have a scale bar in the eye piece, and a 1 mm scale on a microscope slide, called a graticule.

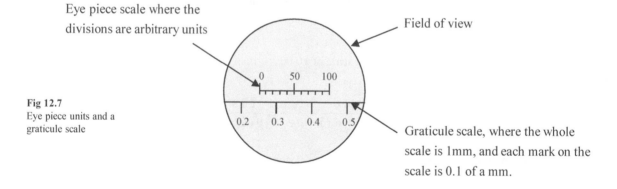

Eye piece scale where the divisions are arbitrary units

Field of view

Fig 12.7
Eye piece units and a graticule scale

Graticule scale, where the whole scale is 1mm, and each mark on the scale is 0.1 of a mm.

You have to ask yourself what proportion of the 1 mm scale does the eye piece cover? What fraction of a millimetre is the eye piece scale? For the scale above the whole eye piece measures 0. 2 mm. Practically you would line the edge of the eye piece scale with the beginning of the graticule scale and read off where the end of the eye piece gets to.

If the eye piece is longer than the graticule, then you can measure a fraction of the eye piece scale against the graticule. For example, half the length, from 0 to 50, on the eye piece scale which goes from 0 to 100 in total, and then in this case you would double the measurement on the graticule for this length of the eye piece to know how many millimetres the eye piece represents.

In the previous figure (Fig 12.7) on the eye piece scale, 50 units is equal to 0.1 mm, so 500 units would be equal to 1 mm. So what ever you measure can be referred to as a proportion of 500 units, and so a proportion of 1mm.

Another way of expressing this would be that the total eye piece scale measures 0.2 mm. So each one unit on the eye piece will be equivalent to 0.002 mm, or 2 μm.

Once you have calibrated the objective of your microscope, you remove the graticule and use the eye piece to determine the size of specimens.

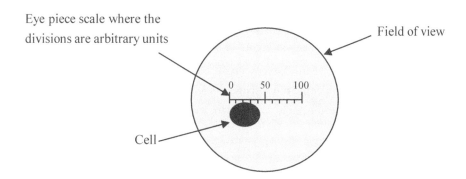

Fig 12.8
Measuring a cell
against eye-piece units

From the calibration, 500 units is equal to 1 mm. So what proportion of 500 units is the specimen?

The specimen needs to be lined up with the beginning of the scale and here the specimen appears to be about 42 units long.
So the question is, what proportion of 500 is 42? Or what fraction?

$$\frac{42}{500} = 0.084$$

Because 500 units is equal to 1 mm, this is the proportion of 1 mm. So the specimen is 0.084 mm, or 84 μm long. So you could say, what proportion of 1 mm is 42 units:

$$\frac{42}{500} \times 1\text{mm} = 0.084 \text{ mm}$$

When ever you make a drawing or other representation of a specimen, such as a photograph, you need to give an indication of the size of the specimen. You could show the measurement of the specimen, showing what its length is, or you could put a scale bar on the page. Here you would measure the length of the specimen in your drawing or photograph and work out what the length is of an appropriate line. For example, if your specimen was 0.084 mm long, or 84 μm, you could display a scale bar that was 10 μm long.

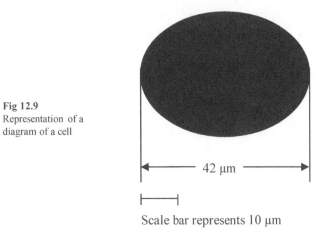

Fig 12.9
Representation of a
diagram of a cell

42 µm

Scale bar represents 10 µm

This drawing shows a cell of 42 µm length in a drawing that has a length of 4.2 cm, or 42 mm.

You could say the diagram is drawn to a scale of 1000:1, meaning 1000 mm on the drawing is 1 mm of the actual cell.

Or that the scale bar, being 1 cm long, represents 10 µm.

Eye piece scales are necessary in microscopy because no microscope is the same as another, and calculating the size of an object that is under the microscope is specific not only to the microscope, but also to the objective you are using. So you need to calibrate the eye piece scale for each objective on each microscope you use.

Another example:
Given the graticule and eye piece measurements below, what would the diameter of the specimen be?

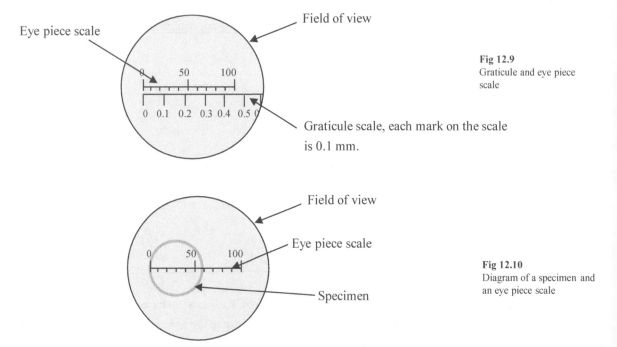

Eye piece scale

Field of view

Fig 12.9
Graticule and eye piece
scale

Graticule scale, each mark on the scale is 0.1 mm.

Field of view

Eye piece scale

Fig 12.10
Diagram of a specimen and
an eye piece scale

Specimen

Eye piece $= \dfrac{0.45}{1}$ x 1 mm = 0.45 mm   so one unit on the eye piece scale $= \dfrac{0.45}{100} = 0.0045$ mm

The specimen measures 60 units on the eye piece, so this is   60 x 0.0045 mm = 0.27 mm.

Test your understanding

1. Ratios

a. You have counted 252 rabbits. You know they have a sex ratio of 4:2 male to female. How many of the rabbits you counted can be assumed to be male?

b. You have 40 patients. 15 are male, the rest are female. Express this as a ratio.

c. In a survey on garden birds, 32 birds were counted. 25 % were robins, 18 % were blackbirds, the rest were wrens. Express this as a ratio.

d. 60% of fish caught in one catch were cod, 20% were plaice and the remainder were halibut. Express this information as a ratio.

e. In an experiment to replicate a field situation, a 4×10^{-4} km^2 field was estimated to have 80% grass, 19.95 % clover and 0.05% dandelion, in terms of the number of plants. There were 96 dandelion plants in total. In the trial plot, a 4 m^2 area was planted with the same % distribution of plants, but in the first year 45% of the grass plants died. How many new grass plants would need to be planted?

2. Percentages

a. In a drug trail, 48% of patients had a severe reaction to a drug. This equated to 1248 patients. How many patients in total took part in the trial?

b. Out of a total of 48 butterflies spotted, 5 were Speckled Wood. Express this as a percentage.

c. Out of 238 children 3 showed symptoms of severe asthma, 14 showed mild symptoms, and the others showed no asthmatic symptoms. Express this data as percentages.

d. In a field of wheat measuring 2.6 hectares, 0.4 hectares was found to be infected with a rust fungus. Express this as a percentage of the total area of wheat.

e. Out of a colony of 640 swans 310 are female. Express this as a decimal.

f. Of an area of tissue stained with a fluorescent dye, 62% of the tissue was found to fluoresce. Express this as a decimal.

3. Dilution

a. A protein solution of unknown concentration was diluted 1/8. A protein assay was performed on the diluted protein sample and it was found to have a concentration of 3 mg ml^{-1}. What was the concentration of the original protein solution?

b. A solution of bromophenol blue had an absorbance reading at 600 nm of 1.2. This solution was diluted 1/5. What do you predict to be the absorbance of the diluted bromophenol blue solution?

c. You are provided with a bacterial suspension with a concentration of 10^5 cells ml^{-1}. By how much would you dilute this suspension to achieve a concentration of 5×10^3 bacteria cells ml^{-1}?

d. You have made 1/2, 1/4 and 1/6 dilutions of a protist culture with 3×10^4 cells ml^{-1}. What is the concentration of protists in each of the dilutions?

e. A 10 mg ml^{-1} solution of DNA was diluted 1/5. What is the new concentration of the DNA in mg ml^{-1}?

f. You need to use 20 pg µl^{-1} of a DNA primer in a PCR reaction. You are supplied with the DNA primer at a concentration of 100 pg µl^{-1}. How much do you need to dilute the DNA primer by to get 20 pg µl^{-1}?

g. Calculate the concentration of bacteria resulting from a 1/6 dilution of a stock of 4 x 10^3 bacteria ml^{-1}. Express this as a percentage of the original concentration, and as a ratio of original:diluted concentrations.

4. Scaling

a. A cell was represented by a diagram with a scale bar of 2 cm representing 2 μm. If the cell in the diagram had a diameter of 12 cm, calculate the 'real life' diameter of the cell.

b. A photograph had a scale of 10000:1. If a chloroplast measured 5 mm on this photograph, what would be the actual measurement of this organelle?

c. If the eye piece on a microscope had a scale of 35 units equal to 1 mm on the graticule slide, and the leg of an arthropod had a diameter of 12 eye piece units, what would the diameter of the arthropod's leg be?

d. A woodland 10 km across is represented on a photograph as being 10 cm across. What is the scale of this photograph?

e. In the question above (no. 4d), if you measured the distance between two oak trees on the photograph as being 15 mm, what is the distance between them in the woodland?

f. Given the eye piece and graticule scales in Fig 12.11, what is the size of the specimen in Fig 12.12?

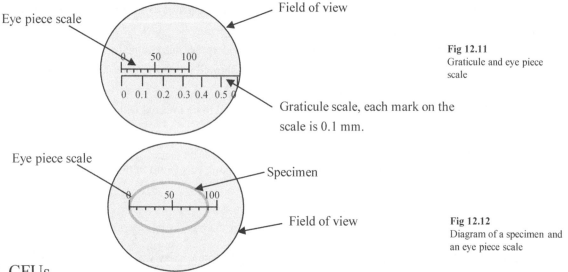

Eye piece scale

Field of view

Fig 12.11
Graticule and eye piece scale

Graticule scale, each mark on the scale is 0.1 mm.

Eye piece scale

Specimen

Field of view

Fig 12.12
Diagram of a specimen and an eye piece scale

5. CFUs

a. You are a soil microbiologist and you want to work out how many culturable bacteria there are in 1 g of soil. The 1 g of soil was placed into 5 ml of sterile water. A serial dilution series was performed from 10^{-1} to 10^{-6}. A sample of 100 μl from each dilution was plated onto culture medium, and incubated at 30°C. After three days the colonies growing on the plates were counted, and 5 colonies were counted at the 10^{-6} dilution. Using this information, calculate the cfu g^{-1} of soil.

b. You are a water microbiologist and you want to work out how many culturable bacteria there are in 100 ml of spring water. A serial dilution series was performed from 10^{-1} to 10^{-6}. A sample of 100 μl from each dilution was plated onto culture medium, and incubated at 30°C. After three days the colonies growing on the plates were counted, and 8 colonies were counted at the 10^{-3} dilution. Using this information, calculate the cfu (100 ml)$^{-1}$ of spring water.

Trigonometry is the study of triangles.

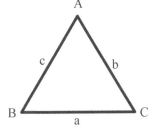

Figure 13.1
A labelled triangle

By custom, triangles are labeled at the corners with capital letters and the corresponding opposite sides are labeled with small letters.

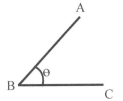

Figure 13.2
A labelled angle

An angle, here given the symbol Θ, can also be referred to as the angle by which two lines, joined at a point, are separated. Here they can be referred to as angle ABC. The size of the angle can be determined by treating the angle as part of a circle.

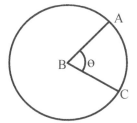

Figure 13.3
A circle showing a sector

The Babylonians, in approximately 2000 BC, decided that a circle could be divided into 360°. They worked in base 60, which is where the number 360 came from. Another way of determining the size of angle Θ is to consider it in terms of properties of the circle.

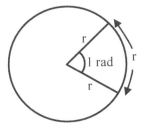

Figure 13.4
A circle showing one radian

If the distance between the two lines, as they reach the perimeter of a circle with radius, r, is equal to the radius, then the angle between the lines is equal to one radian. So the angle is described in terms of a circle, and so is an absolute measurement, rather than something derived.

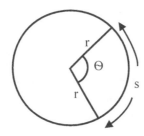

Figure 13.5
A circle showing the
relationship between
an arc and the angle
in radians

If the angle is given the sign Θ, and the length of the arc, s, then when s is equal to the radius, Θ is equal to 1 rad:

$$\Theta = \frac{s}{r}$$

For example, in the diagram below, $r = 4$ cm and $s = 5$ cm.

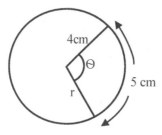

Figure 13.6
Calculating an
angle in radians

$$\Theta = \frac{s}{r} \qquad \Theta = \frac{5 \text{ cm}}{4 \text{ cm}}$$

So here, $\Theta = 1.25$

Note that the value 1.25 has no units. You might see it referred to as 1.25 rad. But if an angle is given with no units, it will be in radians.

If s goes all the way around the circle it is equal to the circumference, so in this case $s = 2\pi r$ because the circumference, $C = 2\pi r$.

From the equation $\theta = \frac{s}{r}$, we know that $s = r\theta$. So we can combine these two equations for s:

$s = 2\pi r$ and $s = r\theta$ to give $r\theta = 2\pi r$. The r on each side cancels out and we are left with $\theta = 2\pi$.

This means that the angle completely round a circle is equal to 2π. This can be useful when converting between radians and degrees, because there are 360 ° around a whole circle.

$$\theta = 2\pi \text{ rad} = 360^o$$

$$\pi \text{ rad} = 180^o$$

Properties of triangles

The sum of the length of two sides is always greater than the length of the third side.
All angles in a triangle add up to 180 °.

Different types of triangle are:
Equilateral triangle – where all sides are equal and all angles = 60° .
Isosceles triangle – two sides are equal and two angles are equal.
Scalene triangle – there are three unequal sides and all angles are different from each other.
Right angled triangle - has one angle of 90 ° and so the other two angles add up to 90 °.

Right-angled triangles

Right angled triangles are useful in biology. For example, is possible to measure the height of objects, knowing your distance from them and the angle from where you are to the top of the object. Also, the angle of movements in joints in arms or legs could be calculated using the rules relating to right angled triangles.

When labelling a right angled triangle the only fixed points are the right angle and the hypotenuse. All other labels depend on the angle you are referring to.

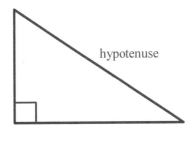

The opposite side, opp, is the side opposite the angle you are referring to.
The adjacent side, adj, is the side adjacent to the angle you are referring to.

Figure 13.7
Labelling the sides
of a triangle
according to the
angle being
considered

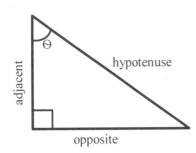

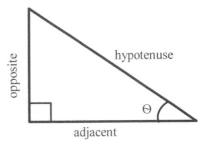

The size and shape of a right angled triangle are defined if the lengths of two sides are known, or one angle and one side. Using the following equations, all properties of a right angled triangle can be calculated once two of its properties are known:

$$\sin\theta = \frac{opp}{hyp} \qquad \cos\theta = \frac{adj}{hyp} \qquad \tan\theta = \frac{opp}{adj}$$

If you know the angle you need to find the sine, cosine or tangent of the angle, as appropriate. If you know two sides, you need to calculate the inverse, $\sin^{-1}$, $\cos^{-1}$ or $\tan^{-1}$ as appropriate. Make sure you can use your calculator to find the following:

$\sin 35° = 0.57$

$\cos 35° = 0.82$

$\tan 35° = 0.70$

$\sin^{-1} 0.25 = 14.48$

$\cos^{-1} 0.25 = 75.52$

$\tan^{-1} 0.25 = 14.04$

Pythagoras

If you know the lengths of two sides you can find the third using Pythagoras' principle:

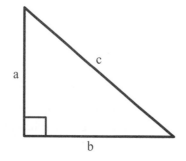

Figure 13.8
A triangle showing sides relating to Pythagoras's Theorum

The square of the hypotenuse is equal to the sum of the squares of the other two sides.

e.g. on the example here $c^2 = a^2 + b^2$

For example:

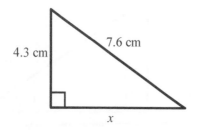

Figure 13.9
A example of Pythagoras's Theorum

Find side x.

$$x^2 = 7.6^2 - 4.3^2$$

$$x^2 = 57.76 - 18.49$$

$$x^2 = 39.27$$

$$x = \sqrt{39.27}$$

$$x = 6.3 \text{ cm}$$

To completely describe a triangle you can use the sine, cosine or tangent equations, Pythagoras' principle, or that all the angles add up to 180 °, one being the right angle, 90 °. It doesn't matter which you use, and in the examples below you could use different combinations of the different methods.

Example 1:

To find side x:

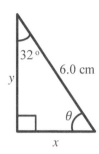

Figure 13.10
Triangle
calculation
example

$$\sin\theta = \frac{opp}{hyp} \qquad opp = \sin\theta \times hyp$$

$$opp = \sin 32 \times 6 \qquad opp = 0.55 \times 6 = 3.179 = 3.2 \text{ cm}$$

Angle $\Theta = 90° - 32°$
Angle $\Theta = 58°$

Side y:
$$y^2 = 6^2 - 3.179^2$$
$$y^2 = 36 - 10.106$$
$$y = \sqrt{25.894} = 5.0886 = 5.1 \text{ cm}$$

Example 2:

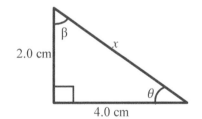

Figure 13.11
Triangle
calculation
example

$$x^2 = 2^2 + 4^2$$

$$x^2 = 4 + 16$$

$$x = \sqrt{20} = 4.47 = 4.5 \text{ cm}$$

$$\tan\theta = \frac{opp}{adj} = \frac{2}{4} \qquad \tan\Theta = 0.5 = 26.565 = 26.6°$$

Angle β: 90 − 26.6 = 63.4 °

Example 3:

Angle Θ: $90 - 63 = 27°$

Side x:

$$\sin 63 = \frac{\text{opp}}{\text{hyp}}$$

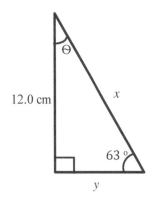

Figure 13.12
Triangle
calculation
example

$$\text{hyp} = \frac{\text{opp}}{\sin 63} = \frac{12}{0.89} = 13.47 = 13.5 \text{ cm}$$

Side y:

$$\cos 63 = \frac{\text{adj}}{\text{hyp}}$$

$$\text{adj} = \cos 63 \times \text{hyp} = 0.45 \times 13.47 = 6.06 = 6.1 \text{ cm}$$

Example 4.

A ramp was set at an angle of 30 ° to the horizontal. A beetle crawled up the ramp for 5 s and travelled 3.8 cm. What height from the ground did the beetle reach?

Figure 13.13
Beetle on a
ramp

The beetle is crawling up the hypotenuse. The question is asking for the height, which in this case would be the side opposite the angle, so:

$$\text{opp} = \text{hyp} \times \sin 30°$$
$$\text{opp} = 3.8\text{cm} \times 0.5 = 1.9\text{cm}$$

The beetle reached a height of 1.9 cm.

How far would the beetle have to travel up the ramp to achieve a height of 6.0 cm?

Now you are being asked to find the hypotenuse, and you are given the side opposite the angle, so:

$$\text{hyp} = \frac{\text{opp}}{\sin 30°} = \frac{6.0}{0.5} = 12.0 \text{ cm}$$

The beetle would have to crawl 12.0 cm.

Example 5

A student stood 20.0 m from the base of a tree and measured the angle from where they stood to the top of the tree to be 57.4 °. What was the height of the tree?

Here, the side adjacent to the angle is the distance from the student to the tree, and you are being asked to calculate the side opposite to the angle. So:

$$\tan \theta = \frac{\text{opp}}{\text{adj}}$$

$$\text{opp} = \tan \theta \times \text{adj}$$

$$\text{opp} = \tan 57.4 \times 20 = 1.564 \times 20 = 31.28$$

The height of the tree was 31.3 m.

Example 6.

A student nurse was instructed to put an arm in a sling, such that the angle at the elbow was 35 ° from the horizontal. She had no protractor. What could she measure to achieve the correct angle?

If she knew the length of the arm she could calculate the height the arm would be needed to be lifted from the horizontal. She measured the arm from the elbow to the tips of the fingers and this was 46 cm. The side she needed to calculate was opposite the angle that needed to be achieved, So:

$$\sin \theta = \frac{\text{opp}}{\text{hyp}}$$

$$\text{opp} = \sin \theta \times \text{hyp}$$

$$\text{opp} = \sin 35 \times 46$$

$$\text{opp} = 0.574 \times 46 = 26.4$$

She would need to lift the arm so that the fingers were 26.4 cm up from the horizontal surface.

Test your understanding

1.   Use the Pythagoras equation to calculate the length of the following:

a.   The hypotenuse where the two other sides are 5.3 mm and 4.8 mm.
b.   The hypotenuse where the two other sides are 6.45 cm and 9.75 cm.
c.   The base of the triangle where the hypotenuse is 45 m and the height is 32 m.
d.   The base where the hypotenuse is 2.4 µm and the height is 1.8 µm.

2. In the following triangle, calculate angle Θ when:

a. $x = 1.2$ cm $y = 4.0$ cm
b. $z = 4.0$ m $x = 5.7$ m
c. $y = 0.48$ mm $z = 0.23$ mm
d. $z = 26$ µm $x = 75$ µm

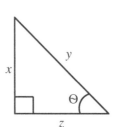

3. In the following triangle, calculate angle Θ when:

a. $x = 4.2$ cm $y = 9.7$ cm
b. $z = 0.3$ m $x = 0.4$ m
c. $y = 2.67$ mm $z = 1.46$ mm
d. $z = 2.1$ µm $x = 3.2$ µm

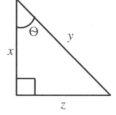

4. In the following triangle calculate y, when:

a.   Θ = 23° $z = 43$ cm
b.   Θ = 41° $x = 0.56$ µm
c.   Θ = 18° $x = 80$ mm
d.   Θ = 38.2° $z = 72.6$ m

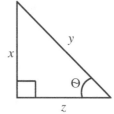

5. In the following triangle calculate z, when:

a.   Θ = 44.3° $x = 72$ µm
b.   Θ = 28.6° $y = 8.79$ µm
c.   Θ = 38° $x = 56$ m
d.   Θ = 41.2° $y = 74.9$ cm

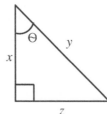

6.A shelter was made by leaning planks against a vertical wall, so that the base of the shelter was 1.4 m and the height of the shelter was 1.6 m.
a.   How long would the planks have to be?
b.   What would the angle be between the foot of the planks and the floor?

What is a mole?

A mole of anything, could be glucose, or salt, or tennis balls, contains 6.022×10^{23} individuals. For glucose it would be 6.022×10^{23} glucose molecules, for salt it would be 6.022×10^{23} salt molecules, for tennis balls it would be 6.022×10^{23} tennis balls. In all these instances, each individual has a mass, in grams. The mass of a salt molecule is extremely tiny, but the mass of 1 mole of salt molecules is 58.44 g. This is a useful amount to work with. A mole of glucose molecules would weigh 180.18 g. A mole of tennis balls would be extremely heavy:

One tennis ball has a mass of 55.15 g

6.022×10^{23} tennis balls have a mass of 3.31×10^{25} g (approximately half the mass of the moon)

The number of individuals in one mole is called Avogadro's number = 6.022×10^{23} individuals mol^{-1}
NB. mol is the abbreviation and the unit of the mole

What is molar mass?

The mass of one mole of anything is referred to as the molar mass and has units of grams per mole. It is the mass of 6.022×10^{23} individuals. The individuals referred to are molecules in the examples below.
The molar mass of NaCl (6.022×10^{23} molecules of NaCl) is 58.44 g mol^{-1}. So to get 1 mol of NaCl you would have to weigh out 58.44g of NaCl.
The molar mass of ZnCl (6.022×10^{23} molecules of ZnCl) is 136 g mol^{-1}. So to get 1 mol of ZnCl you would have to weigh out 136g of ZnCl.

How do you work out the molar mass?

First you have to know the mass of one molecule of the substance you are interested in. To do this you need to know the mass of the atoms making up the molecule. You add together the masses of all the protons and neutrons in the atom (electrons have a negligible mass) and this gives you the mass of the atom. This is going to be an extremely small number. You then add up the masses of the atoms in one molecule. Then this can be multiplied by Avogadro's number to give you the mass of one mole of that molecule.

For example:

Take the carbon atom ^{12}C.   It has 6 protons and 6 neutrons.

The mass of a proton is 1.6726×10^{-24} g

The mass of a neutron is 1.6750×10^{-24} g

So, to take 6 protons and 6 neutrons, you multiply each of the above numbers by 6.

That is: $(1.6726 \times 10^{-24} \text{ g} \times 6) + (1.6750 \times 10^{-24} \text{ g} \times 6)$

you get:

$$10.0356 \times 10^{-24} \text{ g} + 10.0500 \times 10^{-24} \text{ g} = 20.0856 \times 10^{-24} \text{ g}$$

Multiply this by Avogadro's number:

$20.0856 \times 10^{-24} \text{ g} \times 6.022 \times 10^{23} = 12.0955 \text{ g mol}^{-1}$

So the mass in g of a mole of ^{12}C atoms is 12.0955 g mol^{-1}

Or, the molar mass of ^{12}C is 12.0955 g mol^{-1}

NB. Variations can occur depending on how numbers are rounded up in the middle of the equation.

What Avogadro's number does

Avogadro's number converts the very tiny mass of atoms and molecules into a usable number. The number each molecule is converted to, its molar mass, is also approximately equal to the number of protons and neutrons in the molecule. For example, Avogadro's number converts the mass of a mole of ^{12}C to approximately 12 g, and this is very useful because there are 6 neutrons and 6 protons in ^{12}C. This has then been referred to as 12 atomic mass units (amu), equivalent to the number of neutrons and protons in the atom ^{12}C.

The conversion can be applied to the atoms themselves. All atoms and molecules are given a mass relative to one twelfth the mass of ^{12}C, taking the mass of ^{12}C to be 12 amu. This unit is now called the Dalton (Da). By using this conversion, protons and neutrons both have a mass of approximately 1 amu, or 1 Da. This is very convenient because you can work out the number of Da for any molecule, and this amount is equivalent to the number of protons and neutrons in the molecule, and also its molar mass in grams.

Terms you will come across, and what they mean

Dalton
The Dalton is a derived unit, very useful to molecular biologists. You can work out the number of Daltons that make up a molecule, just by adding up the number of protons and neutrons in the molecule, or taking the molar mass and replacing g mol^{-1} with Da.

Molecular mass or molecular weight
Text books use these terms a lot. It means the number of Daltons that make up a molecule. It does not refer to their actual mass in grams, it refers to their mass relative to ^{12}C, as described above. The units of mass cancel out, because it is a relative mass, but it is then assigned the unit of the Dalton

Relative molecular mass (M$_r$)
This is similar to the molecular mass, in that it is the mass of a molecule relative to the mass of one twelfth the mass of ^{12}C. Since ^{12}C is taken to be 12 Da, one twelfth of 12 Da is 1 Da. Because molecular mass is in Daltons, and one twelfth the mass of ^{12}C is 1 Da, the units cancel out. No other units are assigned to this number. So M$_r$ is the same number as molecular mass, but there are no units. Relative molecular mass is what you will find on any container of a chemical (although some list this as Mwt, or molecular weight). The number is equivalent to the molar mass (but molar mass has units of g mol^{-1}), so is a very useful number to know for any chemical.

All these terms and their units are derived. The only expression that is not derived is molar mass. Molar mass is a conversion of atomic mass, using Avogadro's number and has the units of grams per mole. For example:

NaCl   Molar mass = 58.44 g mol^{-1}

Na atomic mass = 22.99 Da
Cl atomic mass = 35.45 Da
Molecular mass = 58.44 Da
Relative molecular mass = 58.44

Glucose C$_6$H$_{12}$O$_6$ Molar mass = 180.18 g mol^{-1}

C atomic mass = 12.01Da
H atomic mass = 1.01 Da
O atomic mass = 16.00 Da
Molecular mass = (12.01 x 6)+(1.01 x 12)+(16.00 x 6) = 180.18 Da
Relative molecular mass = 180.18

Why work in moles?

Knowledge of the number of moles in a compound gives a direct indication of the number of molecules in that compound. For example, if you take 6.022×10^{23} molecules of NaOH and put them with 6.02×10^{23} molecules of HCl, all the molecules will react together. Whereas if you took 1 g of each, there would be molecules of HCl left over because in 1 g NaOH there would be 1.51×10^{22} molecules, and in 1 g of HCl there would be 1.65×10^{22} molecules. So it is the number of moles that is important when looking at chemicals reacting together, because that way you look at the number of molecules of each chemical.

Converting between mass and moles

Molar mass is the mass of a substance equal to one mole, or the mass (g) per mole (mol). So the equation to calculate molar mass is:

$$\text{molar mass (g mol}^{-1}) = \frac{\text{mass (g)}}{\text{moles (mol)}}$$

If this is rearranged to make mass the subject of the equation, then the mass of a substance (g) is equal to the product of the molar mass (g mol^{-1}) and the number of moles (mol).

$$\text{mass (g)} = \text{molar mass (g mol}^{-1}) \times \text{moles (mol)}$$

For example:
Calculate the mass of 0.8 mol ZnCl. The molar mass of ZnCl is 136 g mol^{-1}.

$$\text{mass (g)} = 136 \text{ g mol}^{-1} \times 0.8 \text{ mol}$$

$$\text{mass} = 108.8 \text{ g}$$

To calculate the number of moles in a sample, the equation can be rearranged to give:

$$\text{moles (mol)} = \frac{\text{mass (g)}}{\text{molar mass (g mol}^{-1})}$$

For example:
Calculate the number of moles in 0.6 g glucose. Molar mass of glucose is 180.2 g mol^{-1}.

$$\text{moles (mol)} = \frac{0.6 \text{ g}}{180.2 \text{ g mol}^{-1}}$$

$$\text{moles} = 0.0033 \text{ mol}$$

Further examples:

How many moles of NaCl are there in 0.65 mg of NaCl? Molar mass of NaCl = 58.44 g mol^{-1}.

$$\text{moles (mol)} = \frac{\text{mass (g)}}{\text{molar mass (g mol}^{-1})}$$

$$\text{moles} = \frac{0.65 \times 10^{-3}\text{g}}{58.44 \text{ g mol}^{-1}} = 1.11 \times 10^{-5} \text{ mol}$$

You could also express this answer as 0.0111 mmol or 11.1 μmol

What mass of boric acid is there in 4.56 μmol boric acid? The molar mass of boric acic is 62 g mol^{-1}.

$$\text{mass (g)} = \text{molar mass (g mol}^{-1}) \times \text{moles (mol)}$$

$$\text{mass (g)} = 62 \text{ g mol}^{-1} \times (4.56 \times 10^{-6}) \text{ mol}$$
$$\text{mass} = 2.83 \times 10^{-4} \text{ g}$$

You could also express this answer as 0.283 mg or 283 μg.

Molarity

Molarity is a term used when dealing with solutions. The molarity of a solution refers to the number of moles in one litre, so it is referring to the concentration of a solution. It has the units of mol l^{-1}. So the equation for molarity is:

$$\text{molarity (mol l}^{-1}) = \frac{\text{number of moles (mol)}}{\text{volume (l)}}$$

For example, you could be asked to determine the molarity, or concentration, of a solution with 3 moles of NaCl in 0.5 litres.

Taking the above equation, you would substitute the numbers:

$$\text{molarity (mol l}^{-1}) = \frac{3 \text{ mol}}{0.5 \text{ l}} = 6 \text{ mol l}^{-1}$$

Answer: the molarity of a 0.5 l solution with 3 moles of NaCl is 6 mol l^{-1}

NB. You do not need the molar mass of NaCl in this equation.

Other examples:

What is the molarity, or concentration, of a solution of calcium carbonate which has 0.04 moles calcium carbonate in 10 ml.

$$\text{molarity (mol l}^{-1}) = \frac{\text{number of moles (mol)}}{\text{volume (l)}}$$

The volume must be in litres:

$$\text{molarity (mol l}^{-1}) = \frac{0.04 \text{ mol}}{0.01 \text{ l}} = 4 \text{ mol l}^{-1}$$

What is the molarity of a solution of 0.3 mmol NaCl in 200 ml water?

$$\text{molarity (mol l}^{-1}) = \frac{0.0003 \text{ mol}}{0.2 \text{ l}} = 0.0015 \text{ mol l}^{-1}$$

You could also express this answer as 1.5 mmol l⁻¹.

If you were asked the molarity of a 0.4 l solution with 3 g of NaCl in it (molar mass of NaCl is 58.44 g mol⁻¹), you would use the two previous equations. First identify what you are being asked to find. Here it is the molarity, so you need the equation:

$$\text{molarity (mol l}^{-1}) = \frac{\text{number of moles (mol)}}{\text{volume (l)}}$$

But you do not know the number of moles. So you can use the equation:

$$\text{moles (mol)} = \frac{\text{mass (g)}}{\text{molar mass (g mol}^{-1})} \qquad \text{moles (mol)} = \frac{3 \text{ g}}{58.44 \text{ g mol}^{-1}} = 0.0513347 \text{ mol}$$

Then you can substitute the number of moles into the first equation:
Remember that you do not round up the number in the middle of a calculation.

$$\text{molarity} = \frac{0.0513347 \text{ mol}}{0.4 \text{ l}} = 0.128 \text{ mol l}^{-1}$$

Answer: the molarity of a 0.4 litre solution containing 3 g NaCl is 0.128 mol l⁻¹

Note that units of molarity are sometimes expressed as M, and you will see this in many text books. In this book we use mol l⁻¹ so that units can always be solved in equations. Also, you may see dm³ used instead of litres.

Note that 1 litre = 1 dm³ = 10⁻³ m³.   1 mol l⁻¹ can be expressed as 1 mol dm⁻³ or M

You could answer the previous question by combining the two equations:

$$\text{molarity (mol l}^{-1}) = \frac{\text{number of moles (mol)}}{\text{volume (l)}} \qquad \text{moles (mol)} = \frac{\text{mass (g)}}{\text{molar mass (g mol}^{-1})}$$

Both equations have moles in them, so the first equation can substitute the moles term in the second equation. By substituting the first equation into the number of moles in the second equation you get:

$$\text{molarity (mol l}^{-1}) = \frac{\text{mass (g)}}{\text{molar mass (g mol}^{-1}) \times \text{volume (l)}}$$

Using the previous example to find the molarity of a 0.4 l solution with 3 g of NaCl in it (molar mass of NaCl is 58.44 g mol⁻¹):

$$\text{molarity (mol l}^{-1}) = \frac{3 \text{ g}}{58.44 \text{ g mol}^{-1} \times 0.4 \text{ l}} = 0.128 \text{ mol l}^{-1}$$

Another example:

What mass of NaCl would you need to make 200 ml of a 0.05 mol l⁻¹ solution. Molar mass of NaCl = 58.44 g mol⁻¹.

Using the two equations separately, you are asked to find the mass, so that is the equation

$$\text{mass (g)} = \text{moles (mol)} \times \text{molar mass (g mol}^{-1})$$

But you do not know the number of moles, so you need to first use the equation:

$$\text{moles (mol)} = \text{molarity (mol l}^{-1}) \times \text{volume(l)}$$

$$\text{moles (mol)} = 0.05 \text{ (mol l}^{-1}) \times 0.2 = 0.01 \text{ mol}$$

Then substitute the value for moles into the first equation:

$$\text{mass (g)} = 0.01 \text{ mol} \times 58.44 \text{ g mol}^{-1} = 0.5844 \text{ g}$$

Alternatively you could substitute all the given values into the combined equation which you need to rearrange to make mass the subject:

$$\text{mass (g)} = \text{molarity (mol l}^{-1}) \times \text{molar mass (g mol}^{-1}) \times \text{volume(l)}$$

$$\text{mass (g)} = 0.05 \text{ mol l}^{-1} \times 58.44 \text{ g mol}^{-1} \times 0.2 \text{ l} = 0.5844 \text{ g}$$

Making and diluting solutions

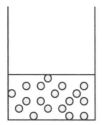

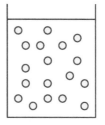

Figure 14.1
Dilution of a substance

If you take a chemical and dilute it, the amount (moles) of chemical before and after dilution is the same, but the concentration of the chemical has changed. Look at the following equation:

Taking the equation:

$$\text{molarity (mol l}^{-1}) = \frac{\text{moles (mol)}}{\text{volume (l)}}$$

This can be rearranged to make moles the subject:

$$\text{moles (mol)} = \text{molarity (mol l}^{-1}) \times \text{volume (l)}$$

If you take the two solutions above the number of moles in each is the same. If the moles stay the same, but the volume has increased, to keep the sides of the equation equal to each other, the molarity, or concentration, must decrease.

If you put a teaspoon of sugar in a cup of tea, and then add more tea, the tea still contains one teaspoon of sugar, but it will taste less sweet. This is because the concentration, or molarity, of the sugar has gone down as the volume goes up.

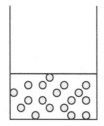

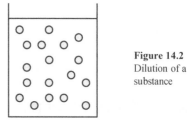

Figure 14.2
Dilution of a substance

$$\text{moles (mol)} = \text{molarity (mol l}^{-1}) \times \text{volume (l)} \qquad \text{moles (mol)} = \text{molarity (mol l}^{-1}) \times \text{volume (l)}$$
$$\text{moles} = M_1 \times V_1 \qquad\qquad\qquad\qquad \text{moles} = M_2 \times V_2$$

Moles are the same for each equation so molarity multiplied by volume must be the same for each equation. Therefore:

$$M_1 V_1 = M_2 V_2$$

So, this equation $M_1V_1 = M_2V_2$ allows you to work out the concentration of a solution before or after it is diluted, or the volume required to make a solution of a specific concentration, or molarity.

Note that you might also see it written as $C_1V_1 = C_2V_2$ where C represents concentration.

Another way of expressing this equation is that the ratio of the first volume to the second volume, equals the ratio of the second concentration over the first.

$$\frac{\text{Initial volume}}{\text{Final volume}} = \frac{\text{Final concentration}}{\text{Initial concentration}}$$

Another version is:

$$\frac{\text{Concentration you've got}}{\text{Concentration you want}} = \frac{\text{volume you want}}{\text{volume you've got}}$$

Check you understand….

Answer the following questions:
1)   What is the difference between moles, molar mass and molarity?
2)   What are the units of molar mass?
3)   What are the units of molarity?
4)   If a solution is diluted what changes, the moles or the molarity?

A note about units

On the whole it is best to work in grams and litres when calculating moles and molarity. If a question is in ml, remember that there are 1000 ml in one litre. It is best to convert units before you start answering the question.

However, with the above equation, it doesn't matter what the units are, as long as they are the same on each side of the equation.

For example:

Your final one litre solution must contain 1.0 mol l^{-1} $(NH_4)_2SO_4$ (ammonium sulphate).   How much 3.0 mol l^{-1} $(NH_4)_2SO_4$ stock must you add?

How to tackle a question like this:

Identify what you have to work out.

The question asks 'How much ….stock must you add?'. The question is asking what volume you must add to get a specific dilution, or what the initial volume is that you have to then dilute.

You need to study the question and identify what information you are given.

Your final one litre solution must contain 1.0 mol l^{-1} $(NH_4)_2SO_4$. How much 3.0 mol l^{-1} $(NH_4)_2SO_4$ stock must you add?

Identifying what you are given:

'Your final one litre solution' – gives you the final volume you require (1 litre)

'3.0 mol l^{-1} $(NH_4)_2SO_4$ stock' – gives you the molarity of the solution you are given (3.0 mol l^{-1})

'must contain 1.0 mol l^{-1}' – gives you the molarity of the solution you want to make (1.0 mol l^{-1})

Working out the equation you need:

When you are taking one solution and making another, you will be using the equation:

$$M_1 V_1 = M_2 V_2$$

There is no mass or molar mass involved, so the other previous equations do not apply here.

Rearranging the equation:

The question asks how much stock you must add, so it wants you to find V_1. The equation must therefore be rearranged by dividing both sides by M_1, and cancelling M_1 from the left hand side of the equation. You end up with:

$$V_1 = \frac{M_2 V_2}{M_1}$$

Alternatively you could say:

$$\text{Volume you must use} = \frac{\text{concentration you want} \times \text{volume you want}}{\text{concentration you have}}$$

Then you plug in the numbers:

$M_1 = 3.0$ mol l^{-1}

$V_1 = ?$

$M_2 = 1.0$ mol l^{-1}

$V_2 = 1$ litre

$$V_1 = \frac{M_2 V_2}{M_1} \qquad\qquad V_1 = \frac{1.0 \times 1}{3.0} = 0.33\,l$$

Check the answer:

Does the answer make sense? You will be adding 0.33 l and making it up to one litre. This means you are diluting you solution by three. Since your initial molarity was 3.0 mol l^{-1} and your final molarity is three times smaller. This answer appears sensible.

A example which uses all the equations relating to moles and molarity:

You are provided with NaCl as a 5 mol l⁻¹ solution, KH_2PO_4 as a solid and glucose as a solution of 550 g l⁻¹.
Molar mass NaCl = 58.4 g mol⁻¹, molar mass KH_2PO_4 = 136.1 g mol⁻¹, molar mass glucose = 180.2 g mol⁻¹.
You have been asked to make a one litre solution that contains a mole of each chemical. How much of each
chemical would you need?

How to tackle a question like this

This question uses all the information you have been given in this chapter. You may panic and wish to scream,
but it is not complicated. All you have to do is:

1.   Split the question up.

2.   For each chemical, work out what information you have and what you need.

3.   Identify the equation or equations for each and plug in the numbers.

4.   Look back and check the answers make sense.

NaCl is provided as a 5 mol l⁻¹ solution. The molar mass of NaCl is 58.44 g mol⁻¹. Your final solution has a
volume of one litre and contains 1 mol NaCl.

This goes from one solution to a second solution, and it is a dilution, so you need:

$$M_1 V_1 = M_2 V_2$$

Remember that molarity is number of moles per litre, so M_2 = 1 mol l⁻¹

$$V_1 = \frac{M_2 V_2}{M_1} = \frac{1 \text{ mol l}^{-1} \times 1 \text{ l}}{5 \text{ mol l}^{-1}} = 0.2 \text{ l}$$

KH_2PO_4 is provided as a solid, so you will want to work out an answer that is in grams.
You have been given the molar mass and you want to find the mass required to give a solution with 1 mole in
one litre.

$$\text{mass (g)} = \text{molarity (mol l}^{-1}) \times \text{molar mass (g mol}^{-1}) \times \text{volume(l)}$$

$$\text{mass (g)} = 1 \text{ mol l}^{-1} \times 136.1 \text{ g mol}^{-1} \times 1 \text{ l} = 136.1 \text{ g}$$

For glucose you have been provided with a solution, and need to find out how much you must dilute it. However, rather than being given the molarity of the initial solution you are given the number of grams in one litre.

The first thing to do is to work out the molarity of the glucose solution you have been given. You can use the molar mass and molarity equations separately, or you can use the combined version, as illustrated here:

$$\text{molarity (mol l}^{-1}) = \frac{\text{mass (g)}}{\text{molar mass (g mol}^{-1}) \times \text{volume (l)}} = 3.05 \text{ mol l}^{-1}$$

So you are given a glucose solution with a molarity of 3.05 mol l⁻¹ and you need to work out how much of this is needed to end up with one mole in one litre. You therefore use the equation:

$$M_1V_1 = M_2V_2$$

You want to know the volume needed to make the dilution, so you need to rearrange the equation to give V_1:

$$V_1 = \frac{M_2V_2}{M_1}$$

Then work out what information you have:

Final molarity, M_2 = 1 mol l⁻¹
Final volume, V_2 = 1 l
Initial molarity, M_1 = 3.05 mol l⁻¹

Then plug in the numbers:

$$V_1 = \frac{1.0 \text{ mol l}^{-1} \times 1 \text{ l}}{3.05 \text{ mol l}^{-1}} = 0.33 \text{ l}$$

So for your three chemicals you would need:

0.2 l of the NaCl solution

136.1 g KH$_2$PO$_4$

0.33 l of the glucose solution

If you were to physically make up this solution you would put the solid chemical and two solutions together, and then make up with water to the desired volume, in this case one litre.

Another example:

You are asked to make up a new media which involves making a 20 ml solution with a final molarity (concentration) of glucose of 0.4 mmol l⁻¹. You have a 100 ml stock solution of glucose and you need to add 1 ml of this stock to 19 ml of the new media. What mass of glucose must the stock contain to ensure you get the right final molarity of glucose in your media. Molar mass of glucose is 180.2 g mol⁻¹.

This is essentially a dilution of one solution to another solution, so the key equation to work with is:

$$M_1V_1 = M_2V_2$$

$M_1 = ?$ It asks you to find the molarity of the stock solution.
$V_1 = 1$ ml   This is what you add from the stock to the new media.
$M_2 = 0.4$ mmol l⁻¹   This is the concentration of the media you are making
$V_2 = 20$ ml   This is the volume of the media you are making

Rearrange the equation to make M_1 the subject:

$$M_1 = \frac{M_2V_2}{V_1}$$

Then plug in the numbers:

$$M_1 = \frac{0.04 \times 20}{1}$$

$$M_1 = 8 \text{ mmol l}^{-1}$$

Note that here you don't need to convert units to mol l⁻¹ and litres because the units of V_1 and V_2, which are in ml, cancel out, and you just have to remember that the units of M_1 are the same as the units of M_2 that you put into the equation.

So you have worked out the molarity of the stock solution you need to make. The question asks you to calculate what mass of glucose you need to achieve this molarity. So now you are working with mass, molarity and molar mass. You can do this using the two separate equations, or the one equation which combines the one for molarity with the one for molar mass. This latter method is the one demonstrated here:

$$\text{molarity (mol l}^{-1}) = \frac{\text{mass (g)}}{\text{molar mass (g mol}^{-1}) \times \text{volume(l)}}$$

You need to rearrange this to make mass the subject:

$$\text{mass (g)} = \text{molarity (mol l}^{-1}) \times \text{molar mass (g mol}^{-1}) \times \text{volume (l)}$$

Note the units in the equation, they are grams, litres and moles, so each of the numbers you put into the equation should be in those units.

M_1 was 8 mmol l⁻¹ so this needs to be expressed as a molarity of 8 x 10⁻³ mol l⁻¹
Volume here is 100 ml – the total volume of the stock solution. This needs to be expressed as 0.1 litres.

$$\text{mass (g)} = 8 \times 10^{-3} \text{ mol l}^{-1} \times 180.2 \text{ g mol}^{-1} \times 0.1 \text{ l}$$

$$\text{mass} = 0.144 \text{ g}$$

So the mass of glucose needed to make the stock is 0.144 g.

Another example:
A technician was asked to make up a solution of potassium iodide with a molarity of 50 μmol l⁻¹, to be used in a practical class where each of 25 groups would require 10 ml of this solution.   The molar mass of KI is 166.0 g mol⁻¹. How much KI would the technician need to weigh out to make up the stock solution?

This is a question involving mass, molarity and molar mass, so you need the equation:

$$\text{molarity (mol l}^{-1}) = \frac{\text{mass (g)}}{\text{molar mass (g mol}^{-1}) \times \text{volume(l)}}$$

but as previously you could use the two separate equations. The question asks about the mass you need to weigh out, so you need to rearrange this formula to make mass the subject:

$$\text{mass (g)} = \text{molarity (mol l}^{-1}) \times \text{molar mass (g mol}^{-1}) \times \text{volume (l)}$$

In the question, the molarity of the solution is 50 μmol l⁻¹. This needs to be converted to mol l⁻¹, so multiply by 10⁻⁶.

For the volume, you have 25 groups, each needing 10 ml, so 250 ml in total. This needs to be converted to litres, so 0.25 l.

$$\text{mass (g)} = 50 \times 10^{-6} \text{ mol l}^{-1} \times 166 \text{ g mol}^{-1} \times 0.25 \text{ l}$$

$$\text{mass} = 2.075 \times 10^{-5} \text{ g}$$

You could express this as 2.075 x 10⁻² mg or 0.02075 μg
So the answer is, that you would need to weigh out 0.02075 μg to make up the stock solution.

Test your understanding.

1. Calculate the following:

a. How many moles of boric acid are there in 3 g boric acid? (Molar mass of boric acid = 62 g mol^{-1})

b. What is the molar mass of KOH if there are 4 moles in 224g of KOH?

c. If you had 0.5 g of KCl in 40 ml, what would the molarity of the KCl solution be? Molar mass of KCl is 74.5 g mol^{-1}.

d. Calculate the molarity of sodium carbonate in a 250 ml solution containing 10 g sodium carbonate. Molar mass of sodium carbonate is 106 g mol^{-1}.

e. How much 0.5 mol l^{-1} solution could you make from 10 ml of 5 mol l^{-1} stock?

f. If you had two stock solutions: 3 mol l^{-1} NaCl and 4 mol l^{-1} KOH, how much of each would you need to make a 30 ml solution of 0.4 mol l^{-1} NaCl and 0.6 mol l^{-1} KOH?

g. Given that there are 0.5 moles of KI in 83 g of KI, what is its molar mass?

h. How much 0.25 mol l^{-1} solution can be made from 500 ml of a 1 mol l^{-1} NaCl stock?

2. Calculate the following:

a. The molar mass of NaCl is 58.44 g mol^{-1}. How many moles of NaCl are there in 4.6 mg?

b. What is the molarity of a 3.5 ml solution that contains 2.92 µg NaCl, with a molar mass of 58.44 g mol^{-1}?

c. How much of a 3.4 mmol l^{-1} solution of KOH would you need to make 200 ml of a 0.25 µmol l^{-1} solution of KOH?

d. The molar mass of glucose is 180 g mol^{-1}. How many moles are there in 35 µg glucose?

e. The molar mass of boric acid is 62 g mol^{-1}. What mass of boric acid is required to give 1.5 mmoles?

f. A 50 ml solution contains 0.78 mg NaCl, with a molar mass of 58.44 g mol^{-1}. What is the molarity of this solution?

g. A 5.5 ml solution, with a molarity of 92 µmol l^{-1} , was made using how much anhydrous sodium acetate, with a molar mass of 82 g mol^{-1}?

h. You have been provided with a 0.25 l solution of NaCl which contains 0.07 moles of NaCl. How much of this would you need to create 30 ml of solution that has a concentration of 4.8 mmol l^{-1} of NaCl?

3. Calculate the following:

a. You have a 1.5 ml sample to which you have added 15 µl of a stock solution of sodium chloride. The stock solution had 3 g of sodium chloride dissolved in 250 ml of water. The molar mass of sodium chloride is 58.44 g mol^{-1}. What is the molarity of sodium chloride in the 1.5 ml sample?

b. A sample of water from the local Broad was tested and found to contain 0.4 mg nitrate per litre. (NO_3 has a molar mass of 62 g mol^{-1}) You need to mimic this in a laboratory. You have a flask that holds 100 ml of solution, and some sodium nitrate with a molar mass of 85 g mol^{-1}. How much sodium nitrate will you need to weigh out so that the molarity of nitrate in the 100 ml solution in the flask matches the molarity of nitrate in the lake?

c. You have been given a sample of lake water which has been analysed and shown to contain 0.4 µg $CuSO_4$ in 1 litre. The level of copper sulphate permitted in the lake is 0.03 µmol l^{-1}. By how much is the sample above or below this permitted level? Molar mass of $CuSO_4$ is 159.6 g mol^{-1}.

Logarithms

What is a logarithm?

A logarithm is a very clever system which takes a number and represents that number as a power to which a particular base is raised to.

For example:

100 can be represented as 10 raised to the power of 2, or $100 = 10^2$

Here the base is 10. Keeping the base as 10, you can see how the number alters as 10 is raised to different powers.

10 raised to the power of 2.4 is 251 or $10^{2.4} = 251$

10 raised to the power of 3.6 is 3981 or $10^{3.6} = 3981$

10 raised to the power of 5.2 is 158489 or $10^{5.2} = 158489$

So you can take the number 100 and rewrite it as 10^2, the number 251 can be rewritten as $10^{2.4}$, the number 3981 can be rewritten as $10^{3.6}$ and the number 158489 can be rewritten as $10^{5.2}$.
Notice that the numbers range from 100 to 158,489 but the powers, to which 10 is raised, range from 2 to 5.2.

Imagine these numbers above are areas of forests, in hectares, representing different sizes of protected habitat. The numbers of jaguars found in each area are 2, 3, 5 and 7 respectively. If you had to represent this on a graph of jaguar numbers against forest area, you would get the following:

Fig 15.1
Jaguar numbers
against forest area

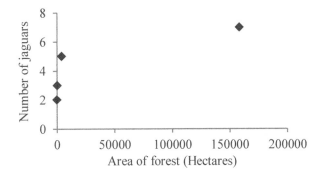

The problem with this graph (Fig 15.1) is that most of the data are squashed into one tiny bit of the graph. Instead of the number of hectares of each area of forest, you could represent this as the number to which 10 is raised to give each area. So 10^2, $10^{2.4}$, $10^{3.6}$ and $10^{5.2}$ could be plotted as 2, 2.4, 3.6 and 5.2:

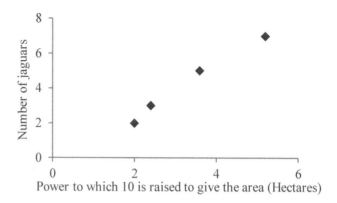

Fig 15.2
Jaguar numbers plotted against the area represented as the power to which 10 is raised.

These powers are referred to as logarithms to the base 10. So by plotting the powers to which 10 is raised, you are actually plotting the logarithms of the numbers. So the *x*-axis of the graph can be referred to as the log of the area in hectares:

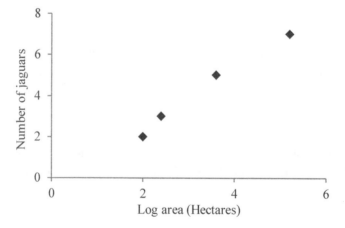

Fig 15.3
Jaguar numbers against the logarithm of the forest area

Using logarithms does two key things to the data. Firstly it can transform data that form a curve into a linear relationship. The reason it changes from a curve to a straight line is that logarithms are used to change a power equation into a linear equation. This can be seen clearly when looking at how you convert between logarithms and exponentials (see below).

Secondly, it spreads the data more evenly along the *x*-axis. This can be useful in many situations. For example, you may need to represent a collection of organisms on an axis, where the organisms range from bacteria to extremely large animals and plants.

	Size	Size in metres	Logarithm of size
Bacterium	0.2 μm	2×10^{-7}	-6.699
Amoeba	0.6 mm	6×10^{-4}	-3.222
Ant	4 mm	4×10^{-3}	-2.398
Mouse	12 cm	1.2×10^{-1}	-0.921
Horse	2 m	2	0.301
Whale	12 m	12	1.079
Giant sequoia tree	120 m	120	2.079

Table 15.1 Standard vs logarithm scale of size of various organisms

From this table you can see clearly that a scale that ranges from the bacterium at 0.2 μm, or 2×10^{-7} m to the giant sequoia tree at 120 m, can be transformed by logarithms to a scale that now ranges from -6.699 to 2.079. These data can then be presented on a scale on a graph or diagram that allows you to make comparisons between very small and very large organisms.

Also note that all the numbers which are less than 1, having a power to which 10 is raised which is a negative number, will have a negative log value

Use the table above to check that you can convert between numbers and logarithms using your calculator.

Logarithms are also very useful in representing a series of powers, for example:

$10^1 = 10$ $\log 10 = 1$
$10^2 = 10 \times 10 = 100$ $\log 100 = 2$
$10^3 = 10 \times 10 \times 10 = 1000$ $\log 1000 = 3$
$10^4 = 10 \times 10 \times 10 \times 10 = 10000$ $\log 10000 = 4$

And so on........

So you can see clearly from this series, how the logarithm is equal to the power to which 10 is raised.

Converting between logarithms and exponentials

Take the expression: $10^{2.4} = 251$

Here the power to which 10 has to be raised, to give 251, is 2.4. You could rephrase this to say that the logarithm to the base 10 of 251 is 2.4.

The logarithm is the power to which the base is raised. However, the base does not have to be 10; it can be any number. Calculators are programmed so that the button labelled 'log' calculates logarithms to the base 10, but this is because it is a base that is often used.

For example: $2^3 = 8$ Here the base is 2, and if 2 is raised to the power 3, it equals 8 (2 x 2 x 2).

This can be rewritten as the logarithm to the base 2 of 8 is 3, or $\log_2 8 = 3$

Another example: $3^4 = 81$ The base is 3 and it is raised to the power 4, which is 3 x 3 x 3 x 3 = 81

Using logarithms it can be written as: $\log_3 81 = 4$ Here the logarithm to the base 3 of 81 is 4.

For any such equation you can use the following:

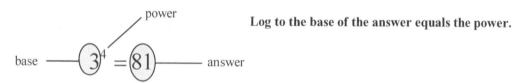

Log to the base of the answer equals the power.

Using this you can transform any exponential equation into a logarithmic equation. When transforming a logarithmic equation into an exponential equation it is useful to remember that the answer of the equation is the power to which the base is raised.

15.1 Laws of logarithms

Logarithms obey several laws, which we now examine. They are introduced via examples.

WORKED EXAMPLE

15.1   Evaluate (a) $\log 7$, (b) $\log 12$, (c) $\log 84$ and $\log 7 + \log 12$. Comment on your findings.

Solution
(a) $\log 7 = 0.8451$ (b) $\log 12 = 1.0792$
(c) $\log 84 = 1.9243$, and $\log 7 + \log 12 = 0.8451 + 1.0792 = 1.9243$

We note that $\log 7 + \log 12 = \log 84$.

Worked Example 15.1 illustrates the **first law** of logarithms, which states:

Key point
$$\log A + \log B = \log AB$$

This law holds true for any base. However, in any one calculation all bases must be the same.

WORKED EXAMPLES

15.2 Simplify to a single log term

(a) $\log 9 + \log x$
(b) $\log t + \log 4t$
(c) $\log 3x^2 + \log 2x$

Solution
(a) $\log 9 + \log x = \log 9x$
(b) $\log t + \log 4t = \log(t.4t) = \log 4t^2$
(c) $\log 3x^2 + \log 2x = \log(3x^2.2x) = \log 6x^3$

15.3 Simplify

(a) $\log 7 + \log 3 + \log 2$
(b) $\log 3x + \log x + \log 4x$

Solution
(a) We know $\log 7 + \log 3 = \log(7 \times 3) = \log 21$, and so

$$\log 7 + \log 3 + \log 2 = \log 21 + \log 2$$
$$= \log(21 \times 2) = \log 42$$

(b) We have

$$\log 3x + \log x = \log(3x.x) = \log 3x^2$$

and so

$$\log 3x + \log x + \log 4x = \log 3x^2 + \log 4x$$
$$= \log(3x^2.4x) = \log 12x^3$$

We now consider an example that introduces the second law of logarithms.

WORKED EXAMPLE

15.4 (a) Evaluate $\log 12$, $\log 4$ and $\log 3$.
   (b) Compare the values of $\log 12 - \log 4$ and $\log 3$.

Solution (a) $\log 12 = 1.0792, \log 4 = 0.6021, \log 3 = 0.4771$.
   (b) From part (a),

$$\log 12 - \log 4 = 1.0792 - 0.6021 = 0.4771$$

and also

$$\log 3 = 0.4771$$

We note that $\log 12 - \log 4 = \log 3$.

This example illustrates the **second law** of logarithms, which states:

Key point

$$\log A - \log B = \log\left(\frac{A}{B}\right)$$

WORKED EXAMPLES

15.5 Use the second law of logarithms to simplify the following to a single log term:

(a) $\log 20 - \log 10$ (b) $\log 500 - \log 75$ (c) $\log 4x^3 - \log 2x$
(d) $\log 5y^3 - \log y$

Solution (a) Using the second law of logarithms we have

$$\log 20 - \log 10 = \log\left(\frac{20}{10}\right) = \log 2$$

(b) $\log 500 - \log 75 = \log\left(\frac{500}{75}\right) = \log\left(\frac{20}{3}\right)$

(c) $\log 4x^3 - \log 2x = \log\left(\frac{4x^3}{2x}\right) = \log 2x^2$

(d) $\log 5y^3 - \log y = \log\left(\frac{5y^3}{y}\right) = \log 5y^2$

15.6   Simplify

(a) $\log 20 + \log 3 - \log 6$
(b) $\log 18 - \log 24 + \log 2$

Solution   (a) Using the first law of logarithms we see that

$$\log 20 + \log 3 = \log 60$$

and so

$$\log 20 + \log 3 - \log 6 = \log 60 - \log 6$$

Using the second law of logarithms we see that

$$\log 60 - \log 6 = \log \left(\frac{60}{6} \right) = \log 10$$

Hence

$$\log 20 + \log 3 - \log 6 = \log 10$$

(b) $\log 18 - \log 24 + \log 2 = \log \left(\dfrac{18}{24} \right) + \log 2$

$$= \log \left(\frac{3}{4} \right) + \log 2$$

$$= \log \left(\frac{3}{4} \times 2 \right)$$

$$= \log 1.5$$

15.7   Simplify

(a) $\log 2 + \log 3x - \log 2x$
(b) $\log 5y^2 + \log 4y - \log 10y^2$

Solution   (a) $\log 2 + \log 3x - \log 2x = \log(2 \times 3x) - \log 2x$

$$= \log 6x - \log 2x$$

$$= \log \left(\frac{6x}{2x} \right) = \log 3$$

(b) $\log 5y^2 + \log 4y - \log 10y^2 = \log(5y^2.4y) - \log 10y^2$

$$= \log 20y^3 - \log 10y^2$$

$$= \log \left(\frac{20y^3}{10y^2} \right) = \log 2y$$

We consider a special case of the second law. Consider $\log A - \log A$. This is clearly 0. However, using the second law we may write

$$\log A - \log A = \log\left(\frac{A}{A}\right) = \log 1$$

Thus

Key point $\log 1 = 0$

In any base, the logarithm of 1 equals 0.
Finally we introduce the third law of logarithms.

WORKED EXAMPLE

15.8 (a) Evaluate $\log 16$ and $\log 2$.
       (b) Compare $\log 16$ and $4 \log 2$.

Solution (a) $\log 16 = 1.204, \log 2 = 0.301$.

       (b) $\log 16 = 1.204, 4 \log 2 = 1.204$. Hence we see that $4 \log 2 = \log 16$.

Noting that $16 = 2^4$, Worked Example 15.8 suggests the **third law** of logarithms:

Key point $n \log A = \log A^n$

This law applies if n is integer, fractional, positive or negative.

WORKED EXAMPLES

15.9 Write the following as a single logarithmic expression:

       (a) $3 \log 2$ (b) $2 \log 3$ (c) $4 \log 3$

Solution (a) $3 \log 2 = \log 2^3 = \log 8$
       (b) $2 \log 3 = \log 3^2 = \log 9$
       (c) $4 \log 3 = \log 3^4 = \log 81$

15.10 Write as a single log term
       (a) $\frac{1}{2}\log 16$ (b) $-\log 4$ (c) $-2 \log 2$ (d) $-\frac{1}{2}\log 0.5$

Solution (a) $\frac{1}{2}\log 16 = \log 16^{\frac{1}{2}} = \log \sqrt{16} = \log 4$

(b) $-\log 4 = -1.\log 4 = \log 4^{-1} = \log\left(\dfrac{1}{4}\right) = \log 0.25$

(c) $-2\log 2 = \log 2^{-2} = \log\left(\dfrac{1}{2^2}\right) = \log\left(\dfrac{1}{4}\right) = \log 0.25$

(d) $-\dfrac{1}{2}\log 0.5 = -\dfrac{1}{2}\log\left(\dfrac{1}{2}\right) = \log\left(\dfrac{1}{2}\right)^{-\frac{1}{2}} = \log 2^{\frac{1}{2}} = \log\sqrt{2}$

15.11 Simplify

(a) $3\log x - \log x^2$
(b) $3\log t^3 - 4\log t^2$
(c) $\log Y - 3\log 2Y + 2\log 4Y$

Solution (a) $3\log x - \log x^2 = \log x^3 - \log x^2$

$$= \log\left(\dfrac{x^3}{x^2}\right)$$

$$= \log x$$

(b) $3\log t^3 - 4\log t^2 = \log(t^3)^3 - \log(t^2)^4$

$$= \log t^9 - \log t^8$$

$$= \log\left(\dfrac{t^9}{t^8}\right)$$

$$= \log t$$

(c) $\log Y - 3\log 2Y + 2\log 4Y = \log Y - \log(2Y)^3 + \log(4Y)^2$

$$= \log Y - \log 8Y^3 + \log 16Y^2$$

$$= \log\left(\dfrac{Y.16Y^2}{8Y^3}\right)$$

$$= \log 2$$

15.12 Simplify

(a) $2\log 3x - \dfrac{1}{2}\log 16x^2$

(b) $\dfrac{3}{2}\log 4x^2 - \log\left(\dfrac{1}{x}\right)$

(c) $2\log\left(\dfrac{2}{x^2}\right) - 3\log\left(\dfrac{2}{x}\right)$

Solution   (a) $2\log 3x - \dfrac{1}{2}\log 16x^2 = \log(3x)^2 - \log(16x^2)^{\frac{1}{2}}$

$$= \log 9x^2 - \log 4x$$

$$= \log\left(\frac{9x^2}{4x}\right)$$

$$= \log\left(\frac{9x}{4}\right)$$

(b) $\dfrac{3}{2}\log 4x^2 - \log\left(\dfrac{1}{x}\right) = \log(4x^2)^{\frac{3}{2}} - \log(x^{-1})$

$$= \log 8x^3 + \log x$$

$$= \log 8x^4$$

(c) $2\log\left(\dfrac{2}{x^2}\right) - 3\log\left(\dfrac{2}{x}\right) = \log\left(\dfrac{2}{x^2}\right)^2 - \log\left(\dfrac{2}{x}\right)^3$

$$= \log\left(\frac{4}{x^4}\right) - \log\left(\frac{8}{x^3}\right)$$

$$= \log\left(\frac{4/x^4}{8/x^3}\right)$$

$$= \log\left(\frac{1}{2x}\right)$$

Self-assessment question 15.1

1. State the three laws of logarithms.

Exercise 15.1

1. Write the following as a single log term using the laws of logarithms:
    (a) $\log 5 + \log 9$ (b) $\log 9 - \log 5$
    (c) $\log 5 - \log 9$ (d) $2\log 5 + \log 1$
    (e) $2\log 4 - 3\log 2$ (f) $\log 64 - 2\log 2$
    (g) $3\log 4 + 2\log 1 + \log 27 - 3\log 12$

2. Simplify as much as possible:
    (a) $\log 3 + \log x$
    (b) $\log 4 + \log 2x$
    (c) $\log 3X - \log 2X$
    (d) $\log T^3 - \log T$
    (e) $\log 5X + \log 2X$

3. Simplify
   (a) $3\log X - \log X^2$ (b) $\log y - 2\log\sqrt{y}$
   (c) $5\log x^2 + 3\log\dfrac{1}{x}$
   (d) $4\log X - 3\log X^2 + \log X^3$
   (e) $3\log y^{1.4} + 2\log y^{0.4} - \log y^{1.2}$

4. Simplify the following as much as possible by using the laws of logarithms:
   (a) $\log 4x - \log x$ (b) $\log t^3 + \log t^4$
   (c) $\log 2t - \log\left(\dfrac{t}{4}\right)$
   (d) $\log 2 + \log\left(\dfrac{3}{x}\right) - \log\left(\dfrac{x}{2}\right)$
   (e) $\log\left(\dfrac{t^2}{3}\right) + \log\left(\dfrac{6}{t}\right) - \log\left(\dfrac{1}{t}\right)$
   (f) $2\log y - \log y^2$
   (g) $3\log\left(\dfrac{1}{t}\right) + \log t^2$
   (h) $4\log\sqrt{x} + 2\log\left(\dfrac{1}{x}\right)$
   (i) $2\log x + 3\log t$ (j) $\log A - \dfrac{1}{2}\log 4A$
   (k) $\dfrac{\log 9x + \log 3x^2}{3}$
   (l) $\log xy + 2\log\left(\dfrac{x}{y}\right) + 3\log\left(\dfrac{y}{x}\right)$
   (m) $\log\left(\dfrac{A}{B}\right) - \log\left(\dfrac{B}{A}\right)$
   (n) $\log\left(\dfrac{2t}{3}\right) + \dfrac{1}{2}\log 9t - \log\left(\dfrac{1}{t}\right)$

5. Express as a single log term:
   $$\log_{10} X + \ln X$$

6. Simplify
   (a) $\log(9x - 3) - \log(3x - 1)$
   (b) $\log(x^2 - 1) - \log(x + 1)$
   (c) $\log(x^2 + 3x) - \log(x + 3)$

Calculating pH

The acidity of a solution is determined by the concentration of H^+ ions. The concentration of these ions ranges from very small (10^{-14} mol l^{-1}) to very large (1 mol l^{-1}). Because the range is so large, logarithms can be used to simplify the values. The equation used is:

$$pH = -\log_{10}[H^+]$$

The minus sign in front of the logarithm means that the pH number is positive, because the H^+ concentration is almost always less than 1. Here are some H^+ concentrations and their pH values.

[H⁺] (mol l⁻¹)	equation	pH
0.1	-log [0.1]	1
10^{-7}	-log [10^{-7}]	7
4×10^{-10}	-log [4×10^{-10}]	9.398

Table 15.2
H⁺ concentrations
and their pH

For *weak* acids and bases, not all of the molecules in the solution ionize to generate H^+ and OH^- ions, much stays in the unionized form, so you cannot take the above pH equation to calculate pH from the concentration of the added acid or base directly. The dissociation constant is a measure of how many of the molecules of acid or base have ionized to generate H^+ or OH^- ions. The dissociation constant for an acid is written as K_a. You can determine K_a experimentally. The higher the value for K_a, the more of the acid has split and hence the more H^+ ions there are. As with pH, values can be very small and very large, so logarithms are taken.

For acids $pK_a = -\log K_a$

For bases $pK_b = -\log K_b$

Buffer systems are composed of an acid-base pair. Each acid-base combination has a specific pH. The pH of an acid-base pair can be calculated based on the formulae

$$pK_a = -\log K_a$$
$$pK_b = -\log K_b$$

A weak acid HA, splits into H^+ and A^- or

$$HA_{(aq)} \leftrightarrow A^-_{(aq)} + H^+_{(aq)}$$

$$K_a = \frac{[A^-][H^+]}{[HA]}$$

or it can be written as

$$K_a = [H^+] \times \frac{[A^-]}{[HA]}$$

Using the first law of logarithms this can be rewritten as

$$\log K_{a=} \log[H^+] + \log[A^-] - \log[HA]$$

You then multiply through the equation by -1:

$$-\log K_{a=} -\log[H^+] - \log[A^-] + \log[HA]$$

This then allows you to substitute the pK_a and pH equations:

$$pK_a = -\log K_a$$
$$pH = -\log_{10}[H^+]$$

into this equation, which gives you:

$$pK_a = pH - \log[A^-] + \log[HA]$$

This equation can then be rearranged further by taking pH across to the other side:

$$-pH + pK_a = -\log[A^-] + \log[HA]$$

And then taking pK_a across to the opposite side:

$$-pH = -pK_a - \log[A^-] + \log[HA]$$

The equation is then multiplied through by -1:

$$pH = pK_a + \log[A^-] - \log[HA]$$

This can be rewritten using the first and second laws of logs, putting the positive logarithm on the top of a fraction, and a negative logarithm on the bottom of a fraction:

$$pH = pK_a + \log\frac{[A^-]}{[HA]}$$

This equation, $pH = pK_a + \log \frac{[A^-]}{[HA]}$, is called the **Henderson-Hasselbalch** equation.

It is worth noting that you don't often calculate pH when making up solutions, because you use a pH meter. However, it is really important to understand the mathematics behind the pH of solutions so that you fully understand what pH is.

Worked examples

[Acetic acid (ethanoic acid) pK_a = 4.7]

1) What would the pH be of an acetic acid/acetate buffer solution containing 0.2 mol l⁻¹ acetic acid (acid or proton donor) and 0.05 mol l⁻¹ sodium acetate (base or proton acceptor)?

$$pH = 4.7 + \log\left(\frac{0.05}{0.2}\right)$$
$$pH = 4.7 - 0.6 = 4.1$$

Answer: The solution would have a pH of 4.1

2) Calculate the pH of a mixture containing 100 ml of 0.1 mol l⁻¹ acetic acid (CH_3CO_2H) and 300 ml of 0.2 mol l⁻¹ sodium acetate ($NaCH_3CO_2$).

In a question like this you must take into account dilution of each solution (100 ml plus 300 ml):

Concentration of acid (acetic acid CH_3CO_2H):

$$\text{no. moles} = 0.1 \times \frac{100}{1000} = 0.01 \text{ moles}$$

Total volume = 400 ml, therefore:

$$0.01 \text{ mol} = \frac{400}{1000} \times \text{molarity}$$

Molarity = 0.025 mol l⁻¹

By same reasoning (but note volumes) concentration of base (sodium acetate $NaCH_3CO_2$):

$$\text{concentration of base} = 0.2 \times \frac{300}{400} = 0.15 \text{mol l}^{-1}$$

Putting these into Henderson-Hasselbalch equation : $pH = pK_a + \log \frac{[A^-]}{[HA]}$

$pH = 4.7 + \log \frac{[0.15]}{[0.025]}$ so, $pH = 4.7 + 0.8 = 5.5$

Answer: The pH would be 5.5.

Natural logarithms

In biology you will generally find only two bases are used:
1) log to the base 10
2) log to the base 2.718281828 ....

Why such a specific number? This number is referred to as '*e*'. It is a number that comes up a lot in mathematics. Historically it is a number that had to be discovered, in the same way that π had to be discovered. The number *e* is linked to numbers that tend to infinity. One of its very early uses was in calculating compound interest.

The number *e*, 2.718281828 .... continues for an infinite number of decimal places. The graph that is related to the number e is the rectangular hyperbola (Fig 13.2). With this graph the values of *x* and *y* continue to infinity. The lines of the graph almost, but never actually, touch the axes.

For this graph when *x* is high *y* is low, and when *y* is high *x* is low, so that $xy = 1$,

or $y = \dfrac{1}{x}$, so where $y = 1$, $x = 1$

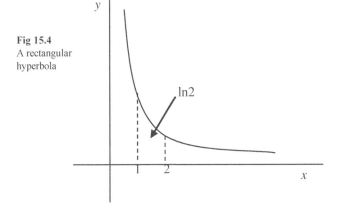

Fig 15.4
A rectangular
hyperbola

This graph has certain properties that are crucial to logarithms and also to calculus.

The area beneath the graph between where $x = 1$ and $x = 2$, is equal to ln 2.

The area under the graph between the points where $x = 1$ and $x = e$ is equal to ln *e* which is equal to 1.

It is the relationship between *x* and *y* on this graph (Fig 15.4) that makes *e* the base of natural logarithms. The key thing to understand is that *e* is a number that relates to numbers that continue to infinity, and that because of this it is a number that occurs a lot in biological processes. The root of *e*, in infinite series, means it is also a really important number in mathematics. Almost all calculations in mathematics and biology that include logarithms use natural logarithms, one exception being calculations of pH.

Calculators are not only programmed to calculate logarithms to the base 10 (log) they are also programmed to calculate logarithms to the base e (ln). Check you can use you calculator correctly for calculating logarithms to the base e:

ln 3 x 10^4 = 10.309
ln 4 x 10^{-5} = -10.127
ln 2.4 = 0.875

You may come across the use of natural logarithms in biology when studying population dynamics, or calculating radioactive decay.

Examples of natural logarithms in biology

The equation used for determining the growth of an exponentially growing population is:
$$N = N_o e^{kt}$$

N = the number in the population
N_0 = the number in the population when t=0
k = growth or decay constant
e = 2.718....
t = time

This can be rearranged to give $\dfrac{N}{N_0} = e^{kt}$ and the using logs it can be transformed to $\ln\dfrac{N}{N_0} = kt$.

You can make an approximation of the growth constant, k, of a population if you know the number that were present at time = 0, and the number after a specific time interval.

For example:
If a population of 30 mice increased to a population of 125 in 160 days, what would the approximate growth constant of this population be?

N_0 = 30 mice t = 160 days
N = 125 mice k = growth constant

$\ln\dfrac{N}{N_0} = kt$ $\ln\dfrac{125}{30} = k \times 160$ $\ln 4.167 = k \times 160$

$1.427 = k \times 160$ $\dfrac{1.427}{160} = k$ $k = 0.009 \; day^{-1}$

Another example:

The numbers of children contracting measles were 756 in 2006. This number is predicted to double within one year. If the number of cases do double in one year, calculate the approximate growth constant for this disease.

$N_0 = 1$ $t = 1$ year

$N = 2$ $k =$ growth constant

$$\ln \frac{N}{N_o} = kt \qquad \ln \frac{2}{1} = k \times 1 \qquad 0.69 = k \times 1 \qquad \frac{0.69}{1} = k \qquad k = 0.69 \, y^{-1}$$

Another example:

The number of parasites per ml of blood in a pigeon rose from 250 to 3600 in three days, what would the approximate growth constant of the parasite population be?

$N_0 = 250$ parasites $t = 3$ days

$N = 3600$ parasites $k =$ growth constant

$$\ln \frac{N}{N_o} = kt \qquad \ln \frac{3600}{250} = k \times 3 \qquad \qquad \ln 14.4 = k \times 3$$

$$2.667 = k \times 3 \qquad \qquad \frac{2.667}{3} = k \qquad k = 0.889 \, day^{-1}$$

Another example where the equation: $N = N_o e^{kt}$ is used in a biological situation, is when calculating decay constants of isotopes. In this case the constant k is referred to as the decay constant, and is a negative number.

For example:

Calculate the decay constant of an isotope with a half life of 6 hours.

For all such questions you can take it that the isotope would give a reading of 1 at time 0, or N_0, and, for the above problem, six hours later the reading, N, would be 0.5, i.e. half what it was six hours earlier.

These values can be substituted into the above equation:

$0.5 = 1e^{k6}$ or $0.5 = e^{6t}$

Logarithms can be used to transform this equation:

$\log_e 0.5 = 6k$ or $\ln 0.5 = 6k$

This can be rearranged to get k on its own: $\frac{\ln 0.5}{6} = k$ $k = -0.116 \, h^{-1}$

k is negative because you are taking the logarithm of number less than one. k is negative if it is a decay constant and positive if it is a growth constant.

Another example:

What is the half life of and isotope with a decay constant of -0.2 day^{-1}?

What is the decay constant of an isotope with a half life of 4 ms?

Test your understanding….

1. Write these exponential expressions in their logarithmic form:

a. $2^4 = 16$

b. $8^3 = 512$

c. $4^3 = 64$

d. $3^5 = 243$

e. $5^2 = 25$

2. Write these logarithmic equations in their exponential form:

a. $\log_2 8 = 3$

b. $\log_4 64 = 3$

c. $\log_7 2401 = 4$

d. $\log_9 81 = 2$

e. $\log_3 243 = 5$

3. Fill in the missing number:

a. $\log_2 ?? = 2$

b. $4^{??} = 16$

c. $\log_{??} 64 = 3$

d. $10^5 = ??$

e. $??^3 = 27$

4. Simplify into a single log term:

a. $\log 6x + \log 4x^2$

b. $\log 4y - \log 2xy$

c. $\log 3x - \log 6x^2 y$

d. $\log 2t + \log 3t^2$

e. $\log 7 + \log 4 - \log 2$

f. $4 \log 2 - 3 \log 2$

g. $3 \log t^2 - 4 \log 2t$

5. Given that acetic acid has a pKa of 4.7, calculate the pH of:

a. an acetic acid/acetate buffer solution containing 0.3 mol l⁻¹ acetic acid (acid or proton donor) and 0.15 mol l⁻¹ sodium acetate (base or proton acceptor).

b. an acetic acid/acetate buffer solution containing 0.1 mol l⁻¹ acetic acid (acid or proton donor) and 0.3 mol l⁻¹ sodium acetate (base or proton acceptor).

c. a mixture containing 200 ml of 0.1 mol l⁻¹ acetic acid (CH_3CO_2H) and 150 ml of 0.2 mol l⁻¹ sodium acetate ($NaCH_3CO_2$)?

d. a mixture containing 100 ml of 0.05 mol l⁻¹ acetic acid (CH_3CO_2H) and 200 ml of 0.3 mol l⁻¹ sodium acetate ($NaCH_3CO_2$)?

6. Calculate the pH of a solution (assume that $[H^+][OH^-] = 10^{-14}$) if:

a.   an acid has a H^+ concentration of 0.00001 mol l^{-1}.

b.   an acid has a H^+ concentration of 0.00000001 mol l^{-1}.

c.   an acid has a H^+ concentration of 0.0005 mol l^{-1}.

d.   a base has an OH^- concentration of 0.007 mol l^{-1}.

e.   a base has an OH^- concentration of 0.03 mol l^{-1}.

f.   a base has an OH^- concentration of 0.02 µmol l^{-1}.

7. Lengths A-E represent a range of lengths. Place these letters on the log scale below according to the length they represent.

A   4 µm

B   7 mm

C   1 m

D   14 m

E   350 km

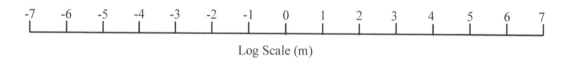

Log Scale (m)

8. The following data comes from a study looking at the number of *Amoeba* in different volumes of soil.

Volume of sample	Number of *Amoebae*
10 µm³	4
1 mm³	20000
400 mm³	6×10^7
18 m³	5×10^{12}

Represent this data on a single graph, using logs.

9. Calculate the following:

a.   the doubling time of a population with an annual growth rate of 4.2%.

b.   the growth rate of a population with a doubling time of 86 years.

c.   the doubling time of a population with an annual growth rate of 0.2 %.

d.   the half life of an isotope with a decay constant of -0.35 day^{-1}.

e.   the decay constant of an isotope with a half life of 30 s.

f.   the half life of and isotope with a decay constant of -0.2 day^{-1}?

g.   the decay constant of an isotope with a half life of 4 ms?

Graphs are a visual representation between two variables, often between a dependent variable and an independent variable. For each graph there is an equation which tells the same story but the graph gives visual impact.

Straight line graphs

Straight line graphs are extremely important and useful to biologists. Key aspects of a straight line graph are the gradient and the intercepts on the x- and y-axes. Straight line graphs are described by the equation $y = mx + c$ where x is the gradient and c is the intercept on the y axis. Many biological problems use straight line graphs to find specific values needed to solve equations and answers to specific questions. For straight line graphs, when the intercept $c = 0$, the value on the x axis is directly proportional to the value on the y axis.

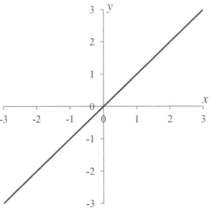

In this case, $y = x$,
so when $y = 2$, $x = 2$,
when $y = -2$, $x = -2$.

Figure 16.1
Graphs of different gradients

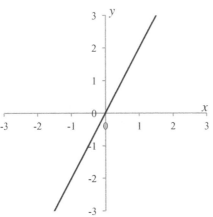

Here, when $y = 2$, $x = 1$
And when $y = -2$, $x = -1$

y is equal to $x \times 2$, or $y = 2x$

The number in front of x is referred to as the gradient. It gives you information about the steepness of the slope of the line and its direction, whether it is positive or negative.

In this graph $y = 2x$. This means that any taken value for x is multiplied by 2 and this gives you the value for y. The graph represents this relationship visually, but the equation $y = 2x$ gives the same information.

Any straight line graph, where $x = 0$ and $y = 0$, has a general equation $y = mx$, where m represents the gradient, or the steepness of the slope.

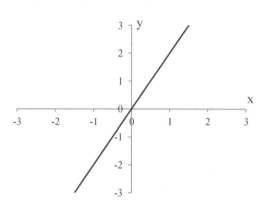

Figure 16.2
$y = 2x$

m = the change in y with respect to x.

Examples of straight-line relationships

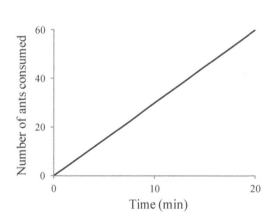

Figure 16.3
Number of ants
consumed over time

y = number of ants consumed
x = time
m = rate at which ants are consumed over time

$$m = \frac{\text{change in } y}{\text{change in } x} = \frac{\Delta y}{\Delta x}$$

When $x = 20$, $y = 60$

$$m = \frac{60 \text{ ants}}{20 \text{ min}} = 3 \text{ ants min}^{-1}$$

The equation of the relationship is $y = 3x$, or the number of ants consumed = 3 x time in minutes
The graph provides a visual representation of this relationship.

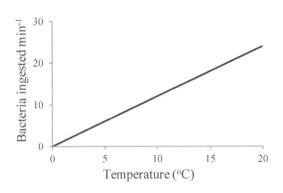

y = number of bacteria ingested per minute (min⁻¹)

x = temperature (°C)

m = rate at which the number of bacteria are ingested as temperature increases

$$m = \frac{\Delta y}{\Delta x} \qquad \text{As before,}$$

$$m = \frac{\text{number of bacteria ingested per minute}}{\text{temperature}}$$

$$m = \frac{24}{20} = 1.2 \text{ min}^{-1}°\text{C}^{-1}$$

Figure 16.4
Effect of temperature
on ingestion of
bacteria

For this relationship there are three ways of describing it:

1) With the graph above.
2) With a written explanation: As the temperature increases, for each ° C, 1.2 more bacteria are ingested per min.
3) With the equation $y = 1.2x$ where y = bacteria ingested min⁻¹ and x = temperature (° C)

Example 1:
A fungal hypha grew according to the equation $y = 5.2x$, where y was the increase in length in mm and x in time in hours. Show this as a graph and as a written description.

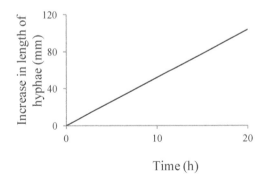

Figure 16.5
Rate of fungal
hyphae growth

The length of the hypha increased in proportion to time. For every hour the fungal hypha will grow 5.2 mm. It is growing at 5.2 mm h⁻¹.

Example 2.
For the following graph provide the equation and description.

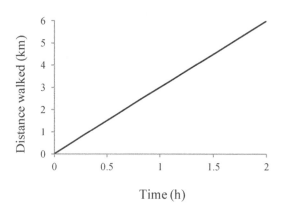

Figure 16.6
Distance walked
over time.

For the equation: $m = \dfrac{6}{2} = 3$ so $y = 3x$ where y = distance walked and x = time (h).

For the description: Distance increased in proportion to time. For every hour the person walked, they covered a distance of 3 km; they were walking at a speed of 3 km h^{-1}.

In all these examples the intercept has been where x and y = zero. Many data sets do not intercept the x- or y-axes at zero. A graph with an intercept on the y axis that is positive looks like this:

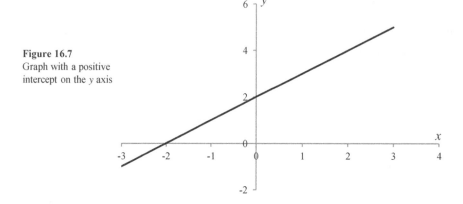

Figure 16.7
Graph with a positive
intercept on the y axis

The generic equation
for this graph is
$$y = mx + c$$
where c is the intercept
on the y-axis.

The equation for this line is $y = 2x + 2$
When $x = 0$, $y = 2$, the point of interception on the y-axis.

When $y = 0$, $0 = 2x + 2$ Solving for x gives you $x = -2$, the point of interception on the x-axis.

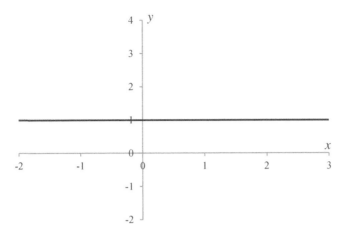

Figure 16.8
Graph with a
zero gradient

Here the gradient is zero, so $y = 0x + c$, and $c = 1$. The equation of the line is therefore $y = 1$.

In the following example the gradient has a negative value, and a negative intercept on the y-axis.

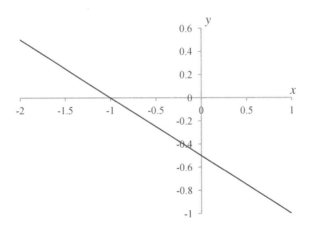

Figure 16.9
Graph with a
negative intercept
on the y axis

The intercept on the y-axis is -0.5. The gradient is negative, y is inversely proportional to time and m has a negative value, in this case $m = $ -0.5, so the equation is $y = -0.5x - 0.5$.

The value for c is unrelated to the slope of the line, it just determines the point where the line crosses the y-axis. The value for m is linked to the value for x, therefore it affects the slope, or gradient, of the line.

The equation of the line tells you everything you need to know about the relationship between x and y. The graph simply illustrates the relationship.

For example, the growth of a tree over time was represented by the equation $y = 2.45x + 0.78$, where y is the growth in metres since planting and x is time in years. The equation tells you that the tree had a height of 0.78 m and grew 2.45 m each year. If you wanted to illustrate this you would need to work out values for x and y so that you can sketch the graph.

Taking this equation, $y = 2.45x + 0.78$, to sketch the graph you need to set some values for x and calculate the respective values for y.

It also might help to calculate the intercept on the x-axis:

x	y
0	0.78
1	3.23
2	5.68
3	8.13

$$0 = 2.45x + 0.78$$
$$-0.78 = 2.45x$$
$$\frac{-0.78}{2.45} = x$$
$$x = -0.3184$$

When $y = 0$, $x = -0.32$, and when $x = 0$, $y = 0.78$. You can then sketch the graph, the key points being the intercepts on the two axes, and that it is a straight line:

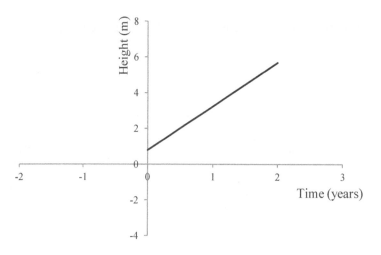

Figure 16.10
Sketching a graph

For example:
Sketch the line on a graph which illustrates the relationship $y = -3x + 2$.

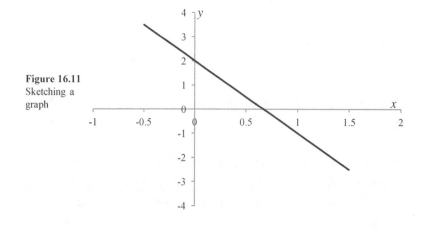

Figure 16.11
Sketching a graph

x	y
0	2
1	-1
2	-4
3	-7

To check the intercept on the x-axis:

$$0 = -3x + 2$$
$$3x = 2$$
$$x = \frac{2}{3}$$

Non-linear graphs in biology

Many relationships in biology are not straight-line relationships. There are ways, however, to transform the data into a straight line relationship. The reason for this is because, with a straight-line relationship, the intercepts on the x- and y-axes can be accurately determined and are often biologically relevant pieces of information.

The exponential graph

This graph follows a line where $y = a^x$

This is the equation for exponential growth where a is a constant. It is applicable, for example, to populations such as bacteria or yeast where individuals divide into two at regular intervals.

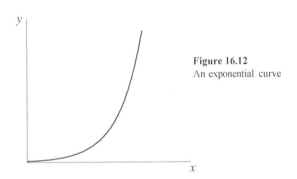

Figure 16.12
An exponential curve

You could sketch a graph of the relationship where the constant was equal to 2, where the equation of the relationship is $y = 2^x$

x	y
0	1
1	2
2	4
3	8
4	16
5	32

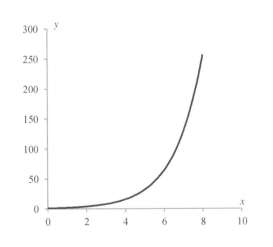

Figure 16.13
Curve where $y = 2^x$

In populations that grow exponentially, this constant is approximately 2.718......, the base of natural logarithms, written as e. A population which follows this exponential growth follows the equation $y = 2.718^x$ or $y = e^x$.

This can be plotted to show the relationship visually:

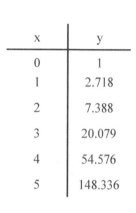

x	y
0	1
1	2.718
2	7.388
3	20.079
4	54.576
5	148.336

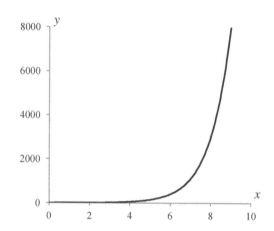

Figure 16.14
Exponential growth

Note that in all these examples when $x = 0$, $y = 1$. This is because anything raised to the power 0 is equal to 1.

A biological situation that uses this relationship is where population growth is calculated. There are two constants, one in front of the base e, which is the number in the population when $x = 0$, and one in front of the power, which is referred to as the growth constant,

The equation used for determining the growth of an exponentially growing population is:

$$N = N_0 e^{kt}$$

N= the number in the population
N_0 = the number in the population when $t = 0$
k = growth or decay constant
e = base of natural logarithms
t = time

This can be plotted on a graph with N on the y-axis and t on the x-axis.

$$N = N_o e^{kt}$$

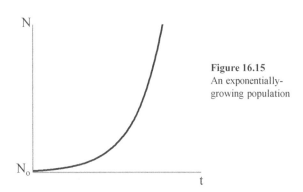

This is an exponential expression. The constant k is a value which can be useful to determine. This is difficult when the relationship is in this form. However, this is an exponential expression and can be converted into a linear expression using logarithms (see chapter on logarithms).

Figure 16.15
An exponentially-growing population

First, the exponential expression needs to be in the form where there is an answer, a base and a power, so that you can rearrange it so it fits the 'log to the base of the answer equals the power'. To do this you move the N_o to the opposite side by dividing both sides by N_o.

$$\frac{N}{N_o} = e^{kt}$$

Here the 'answer' is $\frac{N}{N_o}$, the 'base' is e and the 'power' is kt.

So this is rearranged into its logarithmic form. Logarithms to the base e are natural logarithms, and written as ln:

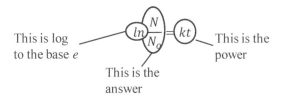

This is log to the base e

This is the answer

This is the power

This is then rearranged using the second law of logarithms, where $\log \frac{A}{B} = \log A - \log B$.

$$\ln N - \ln N_o = kt$$

To plot this on a graph you would plot ln$N$ on the y-axis and t on the x-axis. So lnN_o needs to be moved to the opposite side of the equation by adding lnN_o to both sides:

$$y = mx + c$$

$$\ln N = kt + \ln N_o$$

When this is plotted on a graph of $\ln N$ against t, the intercept on the y-axis is $\ln N_o$ and the gradient is k:

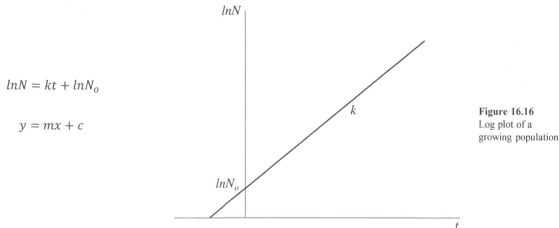

$\ln N = kt + \ln N_o$

$y = mx + c$

Figure 16.16
Log plot of a
growing population

So from this, a value for k can be determined with a high degree of accuracy because it is the gradient of the slope. You could compare the different growth constants of populations under different conditions by comparing the different slopes of the graphs. To look at a dynamic population, however, you need to use calculus, which will be covered in the next chapter.

The hyperbolic, or asymptotic, relationship

You might meet this type of relationship in several different areas of biology. This equation is in the form:
$y = \dfrac{ax}{b+x}$, where a and b are constants.

The line of this relationship curves and reaches its maximum point at infinity. It is at this point that the graph approaches a horizontal, straight line but does not reach it, and this is called an asymptote.

You will come across this type of graph when studying enzyme kinetics. The equation used to describe the action of enzymes on a substrate, or enzyme kinetics is:

$$v = \frac{V_{max}[S]}{K_m + [S]}$$

This is the Michaelis-Menten equation.

Where v = the speed, or velocity, of the enzyme reaction
V_{max} is the line which gives the maximum speed or velocity for v, which the line of the graph never reaches
K_m is the amount of substrate that gives half the maximum rate of reaction for a set amount of enzyme
$[S]$ is the substrate concentration

If this is plotted on a graph of v against $[S]$ you get the following:

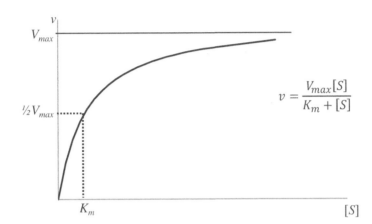

$$v = \frac{V_{max}[S]}{K_m + [S]}$$

Figure 16.17
The hyperbolic curve

To determine the values of the constants V_{max} and K_m, it is much more accurate to convert the relationship into a linear relationship. To do this you need to transpose the equation so that it is in the form $y = mx + c$. There are several ways of doing this, the most commonly used being the Lineweaver-Burk plot.

First the equation is turned upside down.

$$\frac{1}{v} = \frac{K_m + [S]}{V_{max}[S]}$$

The right hand side of the equation can be seen as two fractions over a common denominator, so can be split into two fractions.

$$\frac{1}{v} = \frac{K_m}{V_{max}[S]} + \frac{[S]}{V_{max}[S]}$$

In the second fraction the $[S]$ term cancels out.

$$\frac{1}{v} = \frac{K_m}{V_{max}[S]} + \frac{1}{V_{max}}$$

The first fraction is then split into two separate fractions.

$$\frac{1}{v} = \frac{K_m}{V_{max}} \times \frac{1}{[S]} + \frac{1}{V_{max}}$$

$$y = m \qquad x + c$$

This equation is now in the form recognised as the Lineweaver-Burk plot and is in the form $y = mx + c$, where $\frac{1}{v}$ is plotted on the y-axis and $\frac{1}{[S]}$ on the x-axis.

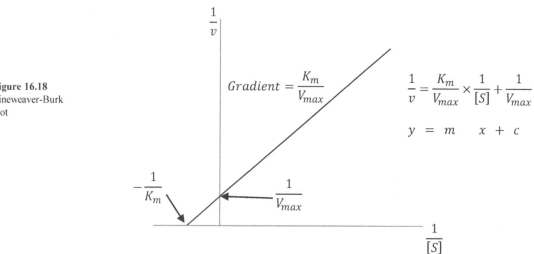

Figure 16.18
Lineweaver-Burk plot

There are two other ways of transposing the Michaelis-Menten equation. The results of this are:

The Eadie-Hofstee plot
$$v = -K_m \frac{v}{[S]} + V_{max}$$

where you get:
$$y = m \ x + c$$

You would plot v against $\frac{v}{[S]}$

and the Hanes-Woolf plot
$$\frac{[S]}{v} = \frac{[S]}{V_{max}} + \frac{K_m}{V_{max}}$$

where you get:
$$y = mx + c$$

You would plot $\frac{[S]}{v}$ against $[S]$.

There are other biological applications of this asymptotic graph. One is in population biology with the Monod equation which relates to microbial growth.

$$\mu = \mu_{max}\frac{s}{K_s + s}$$

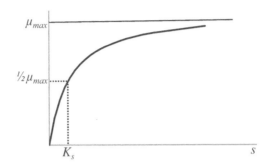

Figure 16.19
The Monod plot

Here μ is the specific growth rate of the population.
μ_{max} is the maximum specific growth rate that the population reaches at infinity.
s is the concentration of a substance that becomes limiting to growth.
K_s is the constant that is the concentration of the substance at half μ_{max}.

It can be rearranged in the same way as the Michaelis-Menten equation, to produce a straight-line relationship.

$$\frac{1}{\mu} = \frac{K_s}{\mu_{max}} \times \frac{1}{s} + \frac{1}{\mu_{max}}$$

$$y = \quad m \quad x + \quad c$$

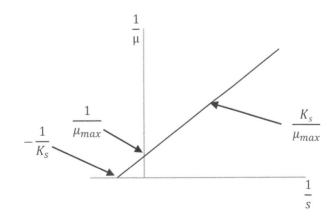

Figure 16.20
The Monod plot
rearranged in the
same way as the
Lineweaver-Burk
plot

Polynomial relationships

Here x is raised to a power. The simplest is the quadratic equation, $y = x^2$. Most quadratic equations follow the equation $y = ax^2 + bx + c$.

If your data follow a line that suggests a quadratic relationship, there will be key points on the line that will be of interest. These are the points where the line crosses the x-axis, the intercept on the y-axis, the maximum or minimum value of y at the point of inflection and the point on the x-axis when y is a maximum or minimum.

If your data follows a line that has an equation:
$y = x^2 + 6x - 8$

It will look like this graph:

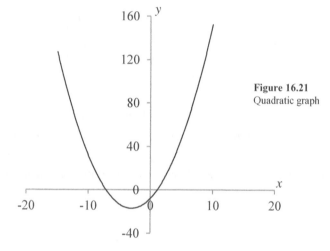

Figure 16.21
Quadratic graph

For a data set from an experiment it is often useful to know the values of x where the line crosses the x-axis, as these may well be biologically important points i.e. where $y = 0$.

To determine this you can put the values from the equation into the following:

$$x = \frac{-b \pm \sqrt{b^2 - 4ac}}{2a}$$

$$y = x^2 + 6x - 8$$
$$y = ax^2 + bx + c$$

so:

$a = 1$ $b = +6$ $c = -8$

The values from the equation for the quadratic graph can be substituted into the equation:

$$x = \frac{-6 \pm \sqrt{6^2 - (4 \times 1 \times -8)}}{2 \times 1}$$

$$x = \frac{-6 \pm \sqrt{36 - (-32)}}{2} \qquad x = \frac{-6 \pm \sqrt{36 + 32}}{2} \qquad x = \frac{-6 \pm 8.24}{2}$$

$$x = -7.12 \quad or + 1.12$$

If you look back at the graph you can see that these relate to the values where the line crosses the x-axis.

If your graph follows the equation $y = x^2 - 6x + 8$ you can find the points where the line crosses the x-axis by factorising the equation. For this equation the graph looks like this:

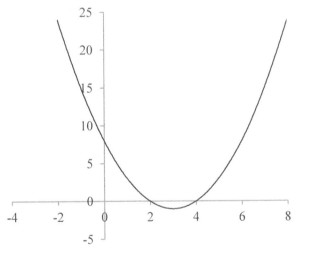

Figure 16.22
A quadratic graph
that factorises
easily

In this case the equation can be factorised to give: $y = (x - 4)(x - 2)$

At the point where the line crosses the x-axis, y = 0, so at this point: $0 = (x - 4)(x - 2)$

For y to equal zero in this equation, one of the terms in brackets must also equal zero, because any value multiplied by zero, will be equal to zero. So you can take each expression in brackets in turn and solve it for x, given that the other expression is equal to zero.

$$0 = 0(x - 2) \quad \text{and} \quad 0 = (x - 4)0$$

Solving each one for x: $x = +2$ and $x = +4$

You can look back at the graph for this relationship and see that these are the points where the line crosses the x-axis.

The minimum point of the curve will be midway between these two points; for this graph it will be where $x = +3$. If this value for x is put into the equation for the graph, you can determine the minimum value of y.

$$y = x^2 - 6x + 8$$
$$y = 3^2 - (6 \times 3) + 8$$
$$y = 9 - 18 + 8$$
$$y = -1$$

Again, you can check this by looking at the point of inflection on the graph – the minimum point of the line.

If there is a minus sign in front of the x^2, the quadratic curve is the other way up – it has a maximum value rather than a minimum value. For example, the line of the relationship $y = -x^2 - 6x + 8$ would be:

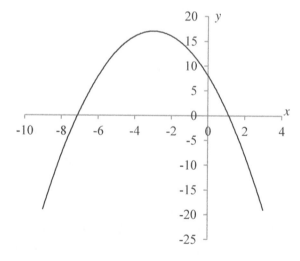

Figure 16.23
Quadratic curve
with a maximum
value

To understand the quadratic equation, it is useful to look at what happens to the shape of graphs as different parts of the equation change. In the previous example you can see what happens when there is a negative sign in front of the x^2 part of the equation. Remember the quadratic equation is the form $y = ax^2 + bx + c$.

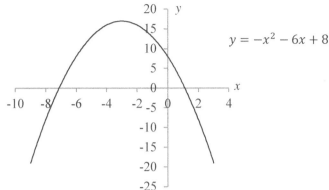

Figure 16.24
A quadratic graph to
compare others to

$$y = -x^2 - 6x + 8$$

Figure 16.25
Example of how quadratic graphs
change with changes in the equation

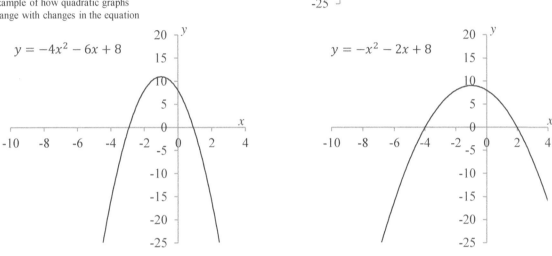

$$y = -4x^2 - 6x + 8$$

$$y = -x^2 - 2x + 8$$

Here the value of a has increased and the gradient of the slope each side of the maximum is steeper.

Here the value of b has increased and it has affected the gradient of the slope and its position on the x axis.

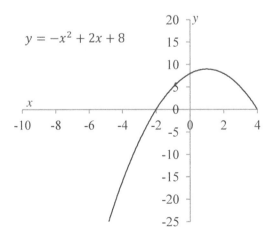

$$y = -x^2 + 2x + 8$$

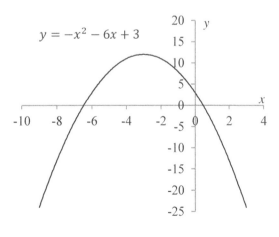

$$y = -x^2 - 6x + 3$$

Here the value of b has increased to a positive number. The line has moved to the right on the x axis and the value for x at the point of inflection is now a positive number.

Here the value of c has changed from +8 to + 3. Shape of the curve remains the same but the point at which it crosses the y axis has changed.

For equations that follow the formula:

$$y = ax^3 + bx^2 + cx + d$$

there are two points of inflection where the line changes direction. Each value in the equation, a, b, c and d, all have different effects on the shape of the curve as they change.

The value d is a constant that is not linked to x. As with the quadratic graph, this value is equal to the point that the curve crosses the y axis.

As with the quadratic graph, each constant affects the shape and position of the graph on the axis. Below are two examples with different values for a and d.

Figure 16.26
Two x^3 graphs

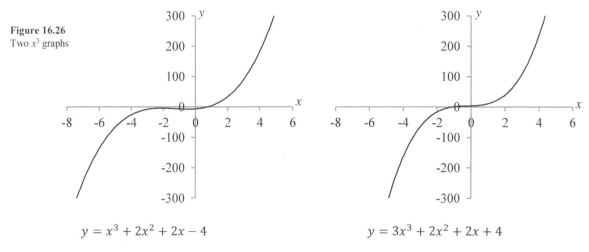

$$y = x^3 + 2x^2 + 2x - 4$$ $$y = 3x^3 + 2x^2 + 2x + 4$$

Here the value for a is 1 and there is a negative intercept, d.

Here the value for a is 3, the gradients either side of the inflection are steeper than the first graph. Also the intercept is positive.

A word of caution here about this type of plot. If you were to produce some data like this below, you can ask a spread sheet software package to put a polynomial regression line through your data.

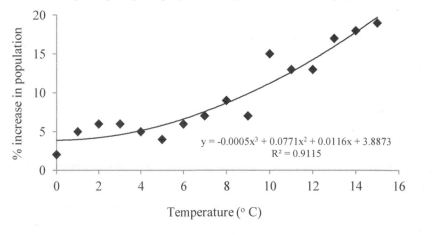

$$y = -0.0005x^3 + 0.0771x^2 + 0.0116x + 3.8873$$
$$R^2 = 0.9115$$

Figure 16.27
A third order polynomial regression line

In the previous graph the R^2 value is 0.9115. The R^2 value tells you how close the points are to the regression line; an indication of the confidence you might have in whether or not the regression line shows a correct relationship between the data. The closer R^2 is to 1, the closer the points are to the chosen regression line.

You can get a higher R^2 value by increasing the order of the polynomial regression. See what happens if you ask the spread sheet software package to fit a fourth order polynomial line to the data.

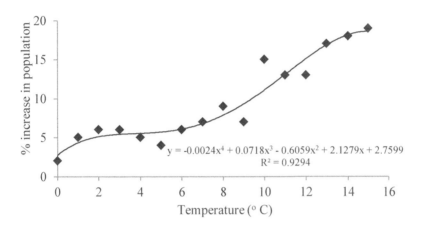

$$y = -0.0024x^4 + 0.0718x^3 - 0.6059x^2 + 2.1279x + 2.7599$$
$$R^2 = 0.9294$$

Figure16.28
A fourth order
polynomial
regression line

The regression line has more points of inflection in it, and fits more closely to the data points, indicated by the R^2 value of 0.9294. Now compare this to fitting a linear regression line through the data.

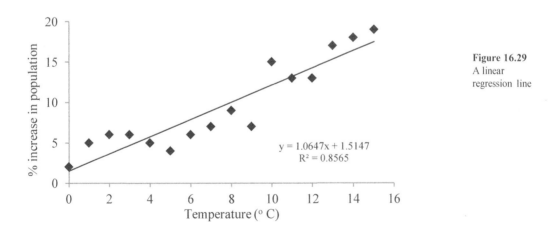

$$y = 1.0647x + 1.5147$$
$$R^2 = 0.8565$$

Figure 16.29
A linear
regression line

The R^2 value might suggest that a polynomial line is the one that best describes the relationship of the data. However, if you use a polynomial plot in this way, you must be able to justify it biologically, not statistically. Many biological situations you will come across will be linear, where the dependent variable is directly proportional to the independent variable, so would merit a straight line regression plot, as above. As with all data, you must think about the biology over and above the statistics.

Test your understanding

1. Sketch a graph for the following equations:
a. $y = 3x - 2$
b. $y = -2x + 2$
c. $y = 3x - 1$
d. $y = 2$
e. $y = 0.45x - 1.2$
f. $y = x$

2. If the growth rate of two cheetah cubs in their first week of life followed the equations:
Cub a) $y = 2.4x + 3$
Cub b) $y = 3.5x + 2$
where y = mass and x = time. Which cub had the faster growth rate? Which cub had the higher birth mass?

3. The Beer Lambert Law uses the equation $A = Ecl$ where:
A is the absorbance, and has no units
E = extinction coefficient (l mol^{-1} cm^{-1})
l = path length (cm), usually 1 cm
c = concentration (mol l^{-1})
a. If E is the gradient, what is plotted on the x- and y-axes?
b. If the graph followed the equation $y = 3.24x$ what would the value of E be, and where would the line of the graph intercept the y axis?

4. From the following graph, which shows how the distance travelled by a tortoise changes with time, what is the speed of the tortoise?

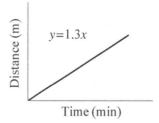

Figure 14.30
Speed of a tortoise

5. For the following equation sketch a graph and describe in words what the relationship is between the dependent and independent variables:
number of seeds germinated = germination rate × time

6. Write the equation and sketch a graph for the relationship where the number of diseased rabbits was found to be directly proportional to the number of rabbits in the population.

7. Sketch graphs for the following equations on a single axis:

a. $y = 1.2^x$

b. $y = 1.4^x$

c. $y = 1.6^x$

8. The concentration of algae in a series of growth chambers was determined every 10 days and their concentration was plotted against time. The equation which represented their growth was found to be

$$y = 0.683e^{0.053x}$$

Using logs, take this equation and convert it to a straight line equation in the form $y = mx + c$ where y is the concentration of algal cells per ml and x is time in days.

9. In the following equations, which are transpositions of the Michaelis-Menten formula, what would you plot on the x and y axes to get straight line plots.

a. $\dfrac{[S]}{v} = \dfrac{[S]}{V_{max}} + \dfrac{K_m}{V_{max}}$

b. $v = -K_m \dfrac{v}{[S]} + V_{max}$

10. From the equation $\ln N = kt + \ln N_0$ what is the gradient in a plot of $\ln N$ against t?

11. In an experiment looking at the growth of *Chara* with increasing copper concentration, the data were plotted and were found to follow a quadratic shape with the equation $y = 0.04x^2 + 4.07x + 12.70$ where y was the growth of *Chara* in µm d^{-1} and x was the concentration of copper in µg l^{-1}.

a.   Does this data have a maximum or a minimum point?

b.   At what concentrations of copper is there no growth of *Chara*?

c.   What is the growth of *Chara* at a copper concentration of 80 µg l^{-1}?

12. Factorise the following equations from quadratic graphs and determine the intercepts on the x axis.

a. $y = x^2 + 3x + 2$

b. $y = x^2 - 8x - 9$

c. $y = x^2 - 36$

d. $y = x^2 - 5x$

e. $y = x^2 + 4x + 4$

13. For the following graphs work out the equation which describes them:

a.

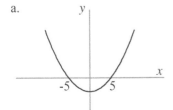

b.

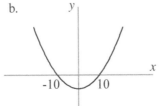

c.

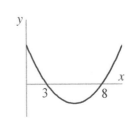

Functions

In an experiment there might be any number of possible variables. Imagine growing a plant. Variables that could be controlled might be temperature (T), nutrients (N), light (L) and water (W). They are all independent of each other; you can change the amount of light and it will not affect the availability of nutrients.

However, the growth of the plant will be affected by any of these variables. So T, N, L and W are all independent variables. Growth (R) is a dependent variable, dependent on values determined for the independent variables.

R depends on T, N, L and W, or you could say R is a function of T, N, L and W. T, N, L and W are all referred to as arguments of the function R.

All of these variables could be kept constant by fixing their values. From an experimental point of view three of the variables should be fixed, so you could, for example, fix N, L and W and test the effect of T on R. The hypothesis would be that plant growth (R) depends on temperature (T). So plant growth would be a function of temperature. This could be written as:

$$R = f(T)$$

You could represent this on a graph, with R on the y-axis, as the dependent variable, and T on the x-axis, as the independent variable.

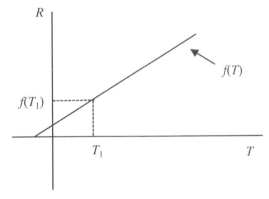

Fig 17.1
Growth as a
function of time

So for a value of T, T_1, there is a value of R which is a function of T, $f(T_1)$. The equation for this relationship is that for a straight line, $y = mx + c$. Here the value for c is the growth of the plant, R, when $T = 0$.

$$f(T) = mT + c$$

For any value of T, there is a value for the growth of the plant which is a function of that temperature.

In general terms, $y = f(x)$.

y is a function of x.

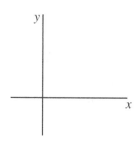

x will have a range of values which you have set.
This is called the domain of the function.
So in the example of plant growth, the range of
temperatures which you grow the plant at is
referred to as the domain of the function.

The values for y (or in the above example, R) which lie within the domain of x (or T) are called the range of the function. So for a range of temperatures, the domain, you get a range of function values.

A function is the result of what is done to x, so   y = f(x):

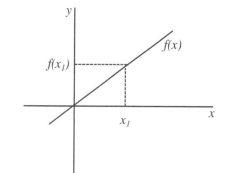

Fig 17.3
Showing the
line f(x)

x_1 has a value which has something done to it to reach a value on the y axis to give $f(x_1)$.

If $y = f(x)$, then for the straight line:

$$f(x) = mx + c$$

$c = 0$ so $f(x) = mx$

m is the operation which is done to x to get a value for f(x).

So if $m = 2$, $f(x) = 2x$, so the operation is 2 x is multiplied by 2 to get to f(x).

Here, 2 is the gradient of the straight line. So in this case, knowing the gradient, means knowing what operation is performed on x to work out what the function of x is. If you know two values for x and f(x) you can work out the gradient.

$$\text{gradient} = \frac{\text{difference in the } y \text{ coordinates}}{\text{difference in the } x \text{ coordinates}}$$

17.1 Finding the equation of a straight line from its graph

If we are given the graph of a straight line it is often necessary to find its equation, $y = mx + c$. This amounts to finding the values of m and c. Finding the vertical intercept is straightforward because we can look directly for the point where the line cuts the y axis. The y coordinate of this point gives the value of c. The gradient m can be determined from knowledge of any two points on the line using the formula

Key point

$$\text{gradient} = \frac{\text{difference between the } y \text{ coordinates}}{\text{difference between the } x \text{ coordinates}}$$

WORKED EXAMPLES

17.1   A straight line graph is shown in Figure 17.4. Determine its equation.

Figure 17.4
Graph for Worked
Example 17.4

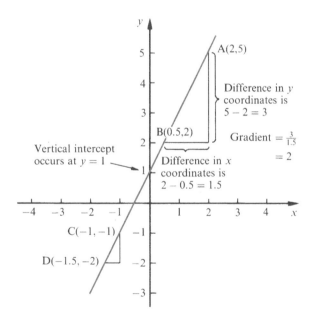

Solution   We require the equation of the line in the form $y = mx + c$. From the graph it is easy to see that the vertical intercept occurs at $y = 1$. Therefore the value of c is 1. To find the gradient m we choose any two points on the line. We have chosen the point A with coordinates (2, 5) and the point B with coordinates (0.5, 2). The difference between their y coordinates is then

$5 - 2 = 3$. The difference between their x coordinates is $2 - 0.5 = 1.5$. Then

$$\text{gradient} = \frac{\text{difference between their } y \text{ coordinates}}{\text{difference between their } x \text{ coordinates}}$$

$$= \frac{3}{1.5} = 2$$

The gradient m is equal to 2. Note that as we move from left to right the line is rising and so the value of m is positive. The equation of the line is then $y = 2x + 1$. There is nothing special about the points A and B. Any two points are sufficient to find m. For example, using the points C with coordinates $(-1, -1)$ and D with coordinates $(-1.5, -2)$ we would find

$$\text{gradient} = \frac{\text{difference between their } y \text{ coordinates}}{\text{difference between their } x \text{ coordinates}}$$

$$= \frac{-1 - (-2)}{-1 - (-1.5)}$$

$$= \frac{1}{0.5} = 2$$

as before.

17.2   A straight line graph is shown in Figure 17.5. Find its equation.

Figure 17.5
Graph for Worked
Example 17.5

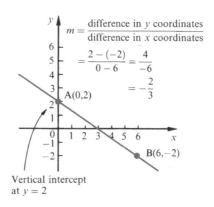

Solution   We need to find the equation in the form $y = mx + c$. From the graph we see immediately that the value of c is 2. To find the gradient we have selected any two points, A(0, 2) and B(6, −2). The difference between their y coordinates is $2 - (-2) = 4$. The difference between their x coordinates is $0 - 6 = -6$. Then

$$\text{gradient} = \frac{\text{difference between their } y \text{ coordinates}}{\text{difference between their } x \text{ coordinates}}$$

$$= \frac{4}{-6}$$

$$= -\frac{2}{3}$$

The equation of the line is therefore $y = -\frac{2}{3}x + 2$. Note in particular that, because the line is sloping downwards as we move from left to right, the gradient is negative. Note also that the coordinates of A and B both satisfy the equation of the line. That is, for A(0, 2),

$$2 = -\frac{2}{3}(0) + 2$$

and for B(6, −2),

$$-2 = -\frac{2}{3}(6) + 2$$

The coordinates of any other point on the line must also satisfy the equation.

The point noted at the end of Worked Example 17.2 is important:

Key point   If the point (a, b) lies on the line $y = mx + c$ then this equation is satisfied by letting $x = a$ and $y = b$.

WORKED EXAMPLE

17.3   Find the equation of the line shown in Figure 17.6.

Figure 17.6
Graph for Worked
Example 17.3

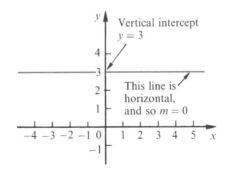

Solution   We are required to express the equation in the form $y = mx + c$. From the graph we notice that the line is horizontal. This means that its gradient is 0, that is $m = 0$. Furthermore the line cuts the vertical axis at $y = 3$ and so the equation of the line is $y = 0x + 3$ or simply $y = 3$.

It is not necessary to sketch a graph in order to find the equation. Consider the following worked examples, which illustrate an algebraic method.

WORKED EXAMPLES

17.4   A straight line passes through A(7, 1) and B(−3, 2). Find its equation.

Solution   The equation must be of the form $y = mx + c$. The gradient of the line can be found from

$$\text{gradient} = m = \frac{\text{difference between their } y \text{ coordinates}}{\text{difference between their } x \text{ coordinates}}$$

$$= \frac{1 - 2}{7 - (-3)}$$

$$= \frac{-1}{10}$$

$$= -0.1$$

Hence $y = -0.1x + c$. We can find c by noting that the line passes through (7, 1), that is the point where $x = 7$ and $y = 1$. Substituting these values into the equation $y = -0.1x + c$ gives

$$1 = -0.1(7) + c$$

so that $c = 1 + 0.7 = 1.7$. Therefore the equation of the line is $y = -0.1x + 1.7$.

17.5   Determine the equation of the line that passes through (4, −1) and has gradient −2.

Solution   Let the equation of the line be $y = mx + c$. We are told that the gradient of the line is −2, that is $m = -2$, and so we have

$$y = -2x + c$$

The point (4, −1) lies on this line: hence when $x = 4$, $y = -1$. These values are substituted into the equation of the line:

$$-1 = -2(4) + c$$

$$c = 7$$

The equation of the line is thus $y = -2x + 7$.

Self-assessment questions 17.1

1. State the formula for finding the gradient of a straight line when two points upon it are known. If the two points are (x_1, y_1) and (x_2, y_2) write down an expression for the gradient.

2. Explain how the value of c in the equation $y = mx + c$ can be found by inspecting the straight line graph.

Exercise 17.1

1. A straight line passes through the two points (1, 7) and (2, 9). Sketch a graph of the line and find its equation.

2. Find the equation of the line that passes through the two points (2, 2) and (3, 8).

3. Find the equation of the line that passes through (8, 2) and (−2, 2).

4. Find the equation of the straight line that has gradient 1 and passes through the origin.

5. Find the equation of the straight line that has gradient −1 and passes through the origin.

6. Find the equation of the straight line passing through (−1, 6) with gradient 2.

7. Which of the following points lie on the line $y = 4x - 3$?
   (a) (1, 2)   (b) (2, 5)   (c) (5, 17)
   (d) (−1, −7)   (e) (0, 2)

8. Find the equation of the straight line passing through (−3, 7) with gradient −1.

9. Determine the equation of the line passing through (−1, −6) that is parallel to the line $y = 3x + 17$.

10. Find the equation of the line with vertical intercept −2 passing through (3, 10).

17.2 Gradients of tangents to curves

Figure 17.7 shows a graph of $y = x^2$. If you study the graph you will notice that as we move from left to right, at some points the y values are decreasing, whereas at others the y values are increasing. It is intuitively obvious that the slope of the curve changes from point to point. At some points, such as A, the curve appears quite steep and falling. At points such as B the curve appears quite steep and rising. Unlike a straight line, the slope of a curve is not fixed but changes as we move from one point to another. A useful way of measuring the slope at any point is to draw a **tangent** to the curve at that point. The tangent is a straight line that just touches the curve at the point of interest. In Figure 17.7 a tangent to the curve $y = x^2$ has been drawn at the point (2, 4). If we calculate the

Figure 17.7
A graph of $y = x^2$

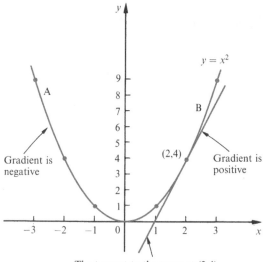

The tangent to the curve at (2,4)

gradient of this tangent, this gives the gradient of the curve at the point (2, 4).

The gradient of a curve at any point is equal to the gradient of the tangent at that point.

WORKED EXAMPLE

17.6 (a) Plot a graph of $y = x^2 - x$ for values of x between -2 and 4.

(b) Draw in tangents at the points $A(-1, 2)$ and $B(3, 6)$.

(c) By calculating the gradients of these tangents find the gradient of the curve at A and at B.

Solution (a) A table of values and the graph are shown in Figure 17.8.

(b) We now draw tangents at A and B. At present, the best we can do is estimate these by eye.

(c) We now calculate the gradient of the tangent at A. We select any two points on the tangent and calculate the difference between their y coordinates and the difference between their x coordinates. We have chosen the points $(-3, 8)$ and $(-2, 5)$. Referring to Figure 17.8 we see that

$$\text{gradient of tangent at A} = \frac{8 - 5}{-3 - (-2)} = \frac{3}{-1}$$

$$= -3$$

Figure 17.8
A graph of $y = x^2 - x$

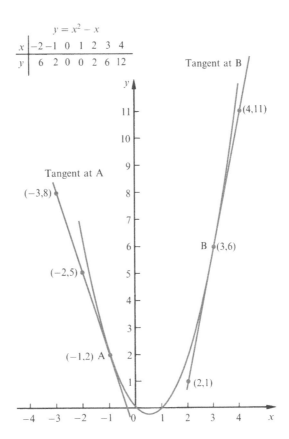

$y = x^2 - x$

x	-2	-1	0	1	2	3	4
y	6	2	0	0	2	6	12

Hence the gradient of the curve at A is -3. Similarly, to find the gradient of the tangent at B we have selected two points on this tangent, namely (4, 11) and (2, 1). We find

$$\text{gradient of tangent at B} = \frac{11 - 1}{4 - 2} = \frac{10}{2}$$

$$= 5$$

Hence the gradient of the tangent at B is 5. Thus the gradient of the curve at B(3, 6) is 5.

Clearly, the accuracy of our answer depends to a great extent upon how well we can draw and measure the gradient of the tangent.

WORKED EXAMPLE

17.7   (a) Sketch a graph of the curve $y = x^3$ for $-2 \leqslant x \leqslant 2$.

(b) Draw the tangent to the graph at the point where $x = 1$.

(c) Estimate the gradient of this tangent and find its equation.

Solution   (a) A graph is shown in Figure 17.9.

Figure 17.9
Graph of $y = x^3$

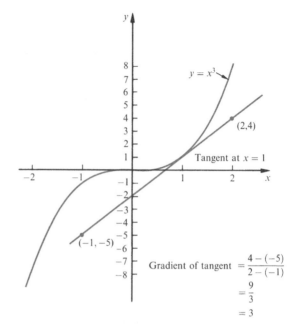

Gradient of tangent $= \dfrac{4 - (-5)}{2 - (-1)}$

$= \dfrac{9}{3}$

$= 3$

(b) The tangent has been drawn at $x = 1$.

(c) Let us write the equation of the tangent as $y = mx + c$. Two points on the tangent have been selected in order to estimate the gradient. These are $(2, 4)$ and $(-1, -5)$. From these we find

$$\text{gradient of tangent is approximately } \frac{4 - (-5)}{2 - (-1)} = \frac{9}{3} = 3$$

Therefore $m = 3$. The value of c is found by noting that the vertical intercept of the tangent is -2. The equation of the tangent is then $y = 3x - 2$.

Of course, this method will usually result in an approximation based upon how well we have drawn the graph and its tangent.

Self-assessment questions 17.2

1. Explain what is meant by the 'tangent' to a curve at a point.

2. Explain how a tangent is used to determine the gradient of a curve.

Exercise 17.2

1. Draw the graph of $y = 2x^2 - 1$ for values of x between -3 and 3. By drawing a tangent, estimate the gradient of the curve at A(2, 7) and B(-1, 1).

2. Draw the graph of $y = -2x^2 + 2$ for values of x between -3 and 3. Draw the tangent at the point where $x = 1$ and calculate its equation.

Test and assignment exercises 17

1. Which of the following will have straight line graphs?

    (a) $y = 2x - 11$ (b) $y = 5x + 10$ (c) $y = x^2 - 1$ (d) $y = -3 + 3x$ (e) $y = \dfrac{2x + 3}{2}$

    For each straight line, identify the gradient and vertical intercept.

2. Find the equation of the straight line that passes through the points (1, 11) and (2, 18). Show that the line also passes through $(-1, -3)$.

3. Find the equation of the line that has gradient -2 and passes through the point (1, 1).

4. Find the equation of the line that passes through $(-1, 5)$ and (1, 5). Does the line also pass through (2, 6)?

5. Draw a graph of $y = -x^2 + 3x$ for values of x between -3 and 3. By drawing in tangents estimate the gradient of the curve at the points $(-2, -10)$ and (1, 2).

6. Find the equations of the lines passing through the origin with gradients (a) -2, (b) -4, (c) 4.

7. Find the equation of the line passing through (4, 10) and parallel to $y = 6x - 3$.

8. Find the equation of the line with vertical intercept 3 and passing through $(-1, 9)$.

9. Find where the line joining $(-2, 4)$ and (3, 10) cuts (a) the x axis, (b) the y axis.

10. A line cuts the x axis at $x = -2$ and the y axis at $y = 3$. Determine the equation of the line.

11. Determine where the line $y = 4x - 1$ cuts
    (a) the y axis   (b) the x axis   (c) the line $y = 2$

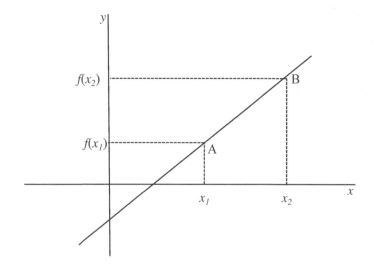

Fig 17.10
Coordinates as
functions

Returning to the idea of functions, you could, for example, give some values, or coordinates, for the figure above for points A and B, for $(x_1, f(x_1))$ and $(x_2, f(x_2))$ (2, 1.2) and (4, 3).

Note: Remember that coordinates first give the value from the x-axis followed by the value from the y axis.

$$\text{gradient} = \frac{\text{difference between } f(x_2) \text{ and } f(x_1)}{\text{difference between } x_2 \text{ and } x_1}$$

$$\text{gradient} = \frac{3 - 1.2}{4 - 2} = \frac{1.8}{2} = 0.9$$

The line is a straight line, so has the equation $f(x) = mx+c$. The gradient is 0.9, so $f(x) = 0.9x+c$.

To find the value for c you can put either values from the coordinates into the equation:

For example:

$3 = (4 \times 0.9) + c$ $1.2 = (2 \times 0.9) + c$

$3 = 3.6 + c$ $1.2 = 1.8 + c$

$3 - 3.6 = c$ $1.2 - 1.8 = c$

$c = -0.6$ $c = -0.6$

Here is an example where you could look at velocity, which is the rate of change of displacement against time. x_1 and x_2 are two different times and $f(x_1)$ and $f(x_2)$ are the respective positions of the object at times x_1 and x_2.

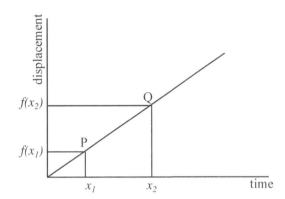

Fig 17.11
Displacement against time
showing functions of x

If you take the above graph, P is the point on the graph where $x = x_1$ and $y = f(x_1)$, and Q is the point where $x = x_2$ and $y = f(x_2)$. To calculate the gradient of the graph you use the equation:

$$\text{gradient} = \frac{\text{difference between } y \text{ coordinates}}{\text{difference between } x \text{ coordinates}}$$

The coordinates of P are $(x_1, f(x_1))$.
The coordinates of Q are $(x_2, f(x_2))$.

$$\text{gradient} = \frac{f(x_2) - f(x_1)}{x_2 - x_1}$$

Rather than using x, the values of x can be defined as different letters, so x_1 can become a, and x_2 can become b:

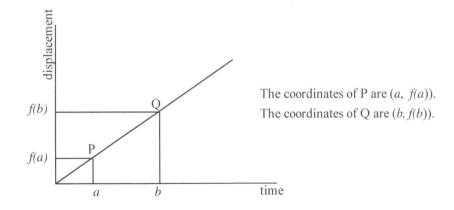

Fig 17.12
Displacement
against time

The coordinates of P are $(a, f(a))$.
The coordinates of Q are $(b, f(b))$.

So the gradient of the line can be calculated from these coordinates:

$$\text{gradient} = \frac{f(b) - f(a)}{b - a}$$

The velocity of an object can therefore be calculated by taking two points in time, knowing the displacement of the object at each of these times.

Now, instead of taking two different times, you can take point b and call it point a plus a length of time, h. In other words: $b = a + h$

Fig 17.13
Showing the
difference
between two
values of x

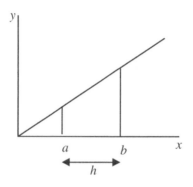

For example, if $a = 3$ min and $b = 5$ min,
then $h = 2$ min
because $b = a + h$
or $5 = 3 + 2$

So the equation from above: $\text{gradient} = \frac{f(b) - f(a)}{b - a}$

can be rewritten substituting $a + h$ for b:

$$\text{gradient} = \frac{f(a + h) - f(a)}{a + h - a}$$

and this can be simplified:

$$\text{gradient} = \frac{f(a + h) - f(a)}{h}$$

This equation can be used to describe the change in position of an object from position $f(a)$ to position $(f\,a+h)$ over a period of time, h. It can also be used on a curve, where you can take a tangent to the curve.

Calculation of gradients of curves

You can get an approximation to the gradient of a curve at a particular point P by taking the secant line (one that intersects at two points on the curve) that intersects at P.

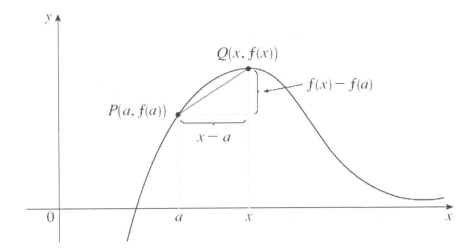

The tangent to the curve $y = f(x)$ at the point P(a, f(a)) is the line through P with a slope which is described by the equation:

$$m = \lim_{x \to a} \frac{f(x) - f(a)}{x - a}$$

This is simply using the equation:

$$\text{gradient} = \frac{\text{difference between the } y \text{ coordinates}}{\text{difference between the } x \text{ coordinates}}$$

The y coordinates are $f(x)$ and $f(a)$, the x coordinates are x and a. In front of this is the term $\lim_{x \to a}$

This term ('lim' refers to limit) refers to the gradient as the value of x approaches the value of a.

Now this curve can be treated in the same way as the straight line has been done previously, by renaming x as $a + h$, where h is the distance between a and x. If you take a situation where the distance h becomes smaller and smaller, x would get closer and closer to a, and so point Q on the curve would move closer to point P. If this happens the secant, or cord, that cuts the curve at two points will cut across a smaller and smaller section of the curve but will still be on the inside of the curve.

As point Q approaches point P, the line that was a secant on the inside of the curve, becomes a tangent on the outside of the curve. The distance h gets smaller and smaller, and the point when the secant becomes a tangent is when the distance approaches zero, referred to as $\lim\limits_{h \to 0}$.

Figure 17.15
As a secant becomes
a tangent

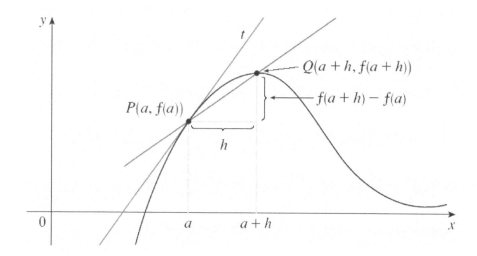

As with the similar treatment of a straight line, substituting $a + h$ for x gives

$$m = \lim_{h \to 0} \frac{f(a+h) - f(a)}{h}$$

The key difference with this equation and the one for a straight line is that this equation needs to be defined as one where h approaches zero, because it refers to a tangent of a curve.

The difference between the treatment of a straight line and a curve can be seen if you look at the difference between average and instantaneous velocities.

Average velocity

Suppose an object moves along a straight line according to an equation of motion $s = f(t)$ where s is the displacement of the object from the origin at time, t.

Using the same process as was used for finding the gradient of a straight line, in the time interval from where $t = a$ to where $t = a + h$, the change in position is $f(a + h) - f(a)$. The average velocity over this time interval is

$$\text{average velocity} = \frac{\text{displacement}}{\text{time}} = \frac{f(a+h) - f(a)}{h}$$

This is the equation for average velocity.

Average and instantaneous velocities

If velocity varies with time, to find the velocity at any particular time it is necessary to draw a tangent to the curve:

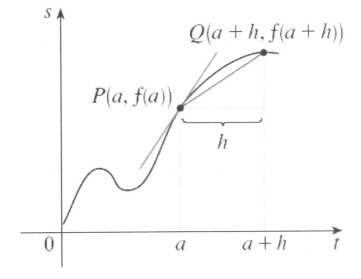

Figure 17.16
Determining
instantaneous
velocities

Average velocity between points P and Q would be $m_{PQ} = \lim_{h \to 0} \dfrac{f(a+h) - f(a)}{h}$

What happens if we compute the average velocities over shorter and shorter time intervals $[a, a + h]$, in other words we let $h \to 0$? We now define the velocity $v(a)$ at time $t = a$ to be the limit of these average velocities:

$$v(a) = \lim_{h \to 0} \frac{f(a+h) - f(a)}{h}$$

This means that the velocity at time $t = a$ (also called the *instantaneous* velocity) is equal to the slope of the tangent line at P.

The difference quotient

Take the equation

$$\text{gradient} = \frac{\text{difference between the } y \text{ coordinates}}{\text{difference between the } x \text{ coordinates}}$$

This is actually saying that the gradient is a change in y over the change in x, or $\frac{\Delta y}{\Delta x}$. This can also be written in the following way, referred to as the difference quotient.

$$\frac{\Delta y}{\Delta x} = \frac{f(x_2) - f(x_1)}{x_2 - x_1}$$

It is called the *average* rate of change of y with respect to x over the interval $[x_1, x_2]$ and can be interpreted as the slope (gradient) of the secant line PQ. As x_2 approaches x_1, $\Delta x \to 0$ and the limit of the average rate of change becomes the (*instantaneous*) rate of change of y with respect to x at $x = x_1$.

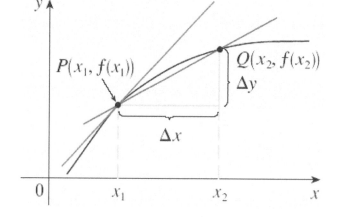

Figure 17.17
The difference
quotient

This is interpreted
as the slope of the
tangent to the curve
$y = f(x)$ at $P(x_1, f(x_1))$

Instantaneous rate of change $= \displaystyle\lim_{\Delta x \to 0} \frac{\Delta y}{\Delta x} = \lim_{x_2 \to x_1} \frac{f(x_2) - f(x_1)}{x_2 - x_1}$

This means that the tangent to the curve at point x_1 becomes equal to the slope of the line joining x_1 and x_2 when the distance between these two points (Δx) approaches zero. This is where, for example, the value of x_2 approaches the value of x_1 and can be calculated from the difference in values for y over the difference in values for x of the slope of the secant, or tangent, to the curve.

Average of change $=$ slope of secant PQ
Instantaneous rate of change $=$ slope of tangent at P

The instantaneous rate of change is physically measurable and is the rate at a particular moment in time. The average rate of change over a particular time period is also a measurable quantity (think average speed cameras!).

Instantaneous change

So the gradient of a line can be described as the difference in the y coordinates divided by the difference in the x coordinates. In a graph of displacement, s, over time, t, the gradient would be the change in displacement, divided by the change in time:

$$gradient = \frac{\Delta s}{\Delta t}$$

If the object was accelerating, the line of the graph would be a curve, and the gradient would be increasing with time.

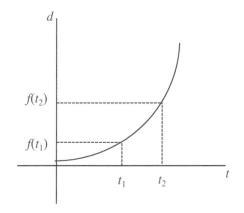

Figure 17.18
Distance
against time

The gradient $= \frac{\Delta s}{\Delta t}$ but this would give an average displacement between t_1 and t_2. You could reduce the length of time between t_1 and t_2, and this would give you a more accurate average displacement, but it would still be an average. If the difference between t_1 and t_2 became extremely small, but was still measureable, you could refer to the gradient as:

$$gradient = \frac{\delta s}{\delta t}$$

t_1 cannot equal t_2, because the difference between t_1 and t_2 cannot be zero. The gradient cannot be a fraction with zero as the denominator. Yet the gradient cannot be determined accurately with a measureable difference between t_1 and t_2. This is where calculus comes in, and until calculus was discovered, independently by Newton and Leibniz in the 17th century, instantaneous change could not be calculated. Calculus opened the door to calculations of things that involved motion, such as the movements of planets and the expansion of gases.

As the difference between t_1 and t_2 approaches zero, calculus allows you to determine how, in this case, displacement changes with respect to time, at a given time. This is given the notation $\frac{ds}{dt}$.

In general terms, for any graph, how y, the dependent variable, changes with respect to x, the independent variable, is given the notation $\frac{dy}{dx}$. This is not a fraction, or ratio in the way that $\frac{\Delta y}{\Delta x}$ or $\frac{\delta y}{\delta x}$ is. It is a symbol that describes the differential.

.

Differential calculus is the mathematical approach to finding the gradient (derivative) of continuous functions. For example: If a population, N, is growing in an exponential fashion with time according to the relationship $N = N_0 e^{rt}$ then the differential of N with respect to time, $\frac{dN}{dt}$, is the gradient to the exponential curve and tells us the rate of change of the population at any given time, t.

For a function of a single variable, $f(x)$, the derivative at a point equals the slope of the tangent line to the graph of the function at that point.

The process of finding a derivative is called differentiation. The fundamental theorem of calculus states that differentiation is the reverse process to integration.

Differentiation has applications in many areas of study. The velocity of a moving body is the derivative of its displacement with respect to time and the derivative of velocity with respect to time is acceleration. The rate of a chemical reaction is a derivative.

Differentiation is frequently used to find the maxima and minima of a function. Equations involving derivatives are called differential equations and are fundamental in describing natural phenomena such as the rate of growth of a population.

To determine how y changes with respect to x, you need to use the Power Rule. The Power Rule is a formula that can be applied to any equation where y is a function of x.

For a given function $y = ax^n$ the Power Rule formula is $\frac{dy}{dx} = anx^{n-1}$ where a is the number in front of x and n is the power to which x is raised.

Take the equation $y = 3x^2$ $a = 3$

$n = 2$

Applying the Power Rule formula:

$$\frac{dy}{dx} = (3 \times 2)x^{2-1} \qquad \text{so,} \qquad \frac{dy}{dx} = 6x$$

Another example:

$$y = 2x^3 \qquad\qquad a = 2$$

$$n = 3$$

Applying the Power Rule:

$$\frac{dy}{dx} = (2 \times 3)x^{3-1} \qquad\qquad \frac{dy}{dx} = 6x^2$$

If you apply the Power Rule to an expression such as $2x$, you could rewrite this as $2x^1$. Similarly an expression such as x^3 could be rewritten as $1x^3$. To apply the Power Rule to a more complex equation, you need to use the Power Rule with each part separately, for example,

$$y = 3x^3 + 2x - 4$$

If you were to rewrite this so that the Power Rule could be applied to each part of this equation it would look like this:
$$y = 3x^3 + 2x^1 - 4x^0$$

Then you can apply the Power Rule:
$$\frac{dy}{dx} = (3 \times 3)x^{3-1} + (2 \times 1)x^{1-1} - (4 \times 0)x^{0-1}$$

and simplify it:
$$\frac{dy}{dx} = 9x^2 + 2x^0 - 0x^{-1}$$
$$\frac{dy}{dx} = 9x^2 + 2$$

Another example:
$$y = x^2 - 4x - 7$$

This can be rewritten as
$$y = 1x^2 - 4x^1 - 7x^0$$

Then you apply the Power Rule:
$$\frac{dy}{dx} = (2 \times 1)x^{2-1} - (4 \times 1)x^{1-1} - (7 \times 0)x^{0-1}$$

and simplify it:
$$\frac{dy}{dx} = 2x^1 - 4x^0 - 0x^{-1}$$

$$\frac{dy}{dx} = 2x - 4$$

So you can see that any number that is not associated with x disappears because x will be to the power zero, and any number multiplied by zero, equals zero. The differential is looking at how y varies with respect to x, and a number on its own, not associated with x, will not affect this relationship, and do disappears when you apply the Power Rule. Also, for any number associated with x^1 , the x disappears. So when applying the Power Rule to any equation you can effectively ignore a number on its own and remove the x from an expression where x is to the power 1.

Proving the Power Rule

Here is an example to show how the Power Rule can be proved. Let's look at differentiation of the function,

$$y = x^2$$

If the Power Rule is applied to this expression, you will get

$$\frac{dy}{dx} = 2x$$

But what does the $\frac{dy}{dx}$ mean?

We can predict the effect of making a small change, δy, to y and a small change δx, to x.

$$y + \delta y = (x + \delta x)^2$$

Expanding the contents of the brackets gives
$$y + \delta y = x^2 + 2x\delta x + \delta x^2$$

Now removing the equality $y = x^2$ gives the expression for the small change in x and y, rather than values of x and y.
$$\delta y = 2x\delta x + \delta x^2$$

To get the expression for a change in y over a change in x on one side of the equation, both sides of the equation are divided by δx
$$\frac{\delta y}{\delta x} = \frac{2x\delta x}{\delta x} + \frac{\delta x^2}{\delta x}$$

and simplified

$$\frac{\delta y}{\delta x} = 2x + \delta x$$

Finally, to derive the *instantaneous* rate of change of y with respect to x, we need to make the small change in x, δx vanishingly small i.e. take the limit $\delta x \to 0$. Under these circumstances, the average rate of change $\delta y/\delta x$ becomes the instantaneous rate and the way it is written changes to dy/dx to denote this. This is the derivative which we are seeking

$$\frac{dy}{dx} = 2x$$

Take the equation $y = 2x^{-1}$. If you were to apply the Power Rule to this equation you would follow exactly the same process as with the previous expressions.

$$\frac{dy}{dx} = anx^{n-1}$$

Here, n = -1, so applying the Power Rule,

$$an = 2 \times -1$$

and

$$n - 1 = -1 - 1.$$

so

$$\frac{dy}{dx} = -2x^{-2}$$

This can also be written as

$$\frac{dy}{dx} = -2\frac{1}{x^2}$$

For example:

Differentiate $y = 3x^{-2}$ with respect to x.

$$\frac{dy}{dx} = (3 \times -2)x^{-2-1}$$

$$\frac{dy}{dx} = -6x^{-3}$$

Another example:

Differentiate $y = 4x^{-1} + 2x^2$ with respect to x.

$$\frac{dy}{dx} = (4 \times -1)x^{-1-1} + (2 \times 2)x^{2-1}$$

$$\frac{dy}{dx} = -4x^{-2} + 4x$$

Another example:

Differentiate $y = 3t^{-3} - t^2 + 45$ with respect to t. This is exactly the same as the other examples, except that x is replaced by t, which might be, for example, is a graph with time on the x axis.

$$\frac{dy}{dx} = (3 \times -3)t^{-3-1} - (1 \times 2)t^{2-1} + (45 \times 0)t^{0-1}$$

$$\frac{dy}{dx} = -9t^{-4} - 2t$$

The Power Rule is a formula that can be applied to any equation where y is a function of x. For example, you can use the Power Rule to calculate the value of x at the point at which the gradient of a curve is equal to zero, at the point of inflection.

Take this graph:

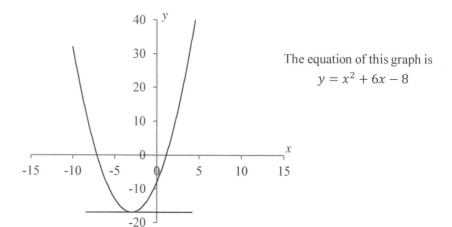

The equation of this graph is
$$y = x^2 + 6x - 8$$

Fig 17.19
Applying the Power Rule
to the minimum value of a
quadratic curve

To apply the Power Rule, $\frac{dy}{dx} = anx^{n-1}$, to the equation of the line, first the equation can be rewritten as:

$$y = 1x^2 + 6x^1 - 8x^0$$

Then:

$$\frac{dy}{dx} = (1 \times 2)x^{2-1} + (6 \times 1)x^{1-1} - (8 \times 0)x^{0-1}$$

Simplifying this:

$$\frac{dy}{dx} = 2x^1 + 6x^0 - 0x^{-1}$$

$$\frac{dy}{dx} = 2x + 6$$

This has determined the equation for $\frac{dy}{dx}$, the instantaneous change in y with respect to x. To apply this to the point of inflection of the curve you can see that at this specific point the rate of change of y with respect to x is equal to zero, that is, $\frac{dy}{dx} = 0$.

So if $\frac{dy}{dx} = 0$, at this specific point on the curve, where $\frac{dy}{dx} = 2x + 6$, $2x + 6 = 0$.

You can therefore solve the equation for x:

$2x = -6$ $x = \frac{-6}{2}$ $x = -3$

This gives you the value for x where the line reaches its minimum value. If you wanted to find the value for y at this point, you would substitute the value for x or -3 into the original equation:

$$y = x^2 + 6x - 8$$
$$y = (-3)^2 + (6 \times -3) - 8$$
$$y = 9 - 18 - 8$$
$$y = -17$$

How can this process be applied to a biological situation?

Take, for example, a study to find the isoelectric point (pI) of the protein casein, where the following data was obtained, and a graph plotted of pH against protein concentration (Fig 17.20). Note that the solubility of a protein is least when the pH equals its pI. So, if we can find the pH at which the protein concentration is minimized, then this pH will be equal to the pI.

pH	Protein concentration (mg ml^{-1})
5.6	0.46
5.3	0.35
5.0	0.25
4.7	0.14
4.4	0.10
4.1	0.15
3.8	0.30
3.5	0.45

If you plot this data and estimate the minimum point by eye you would probably estimate a value for the protein concentration of around 0.1 mg ml^{-1} at a pH of around 4.4.

Given that data such as this is determined with as much experimental accuracy as possible, to then estimate the minimum point would not reflect the accuracy of the experiment. To determine the minimum point mathematically provides this accuracy.

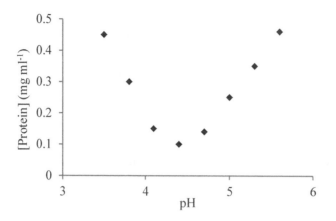

Fig 17.20
The isoelectric
point of casein

Alternatively you could enter the data into a spread sheet software package and ask the programme to add a quadratic line to the data (polynomial 2, or x^2). You also need the programme to display the equation on the graph.

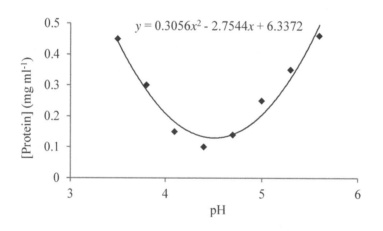

$$y = 0.3056x^2 - 2.7544x + 6.3372$$

Fig 17.21
Using an equation of
the curve to find the
isoelectric point of
casein

You then apply the Power Rule to this equation :

$$y = 0.3056x^2 - 2.7544x + 6.3372,$$

as before:

$$y = 0.3056x^2 - 2.7544x^1 + 6.3372x^0$$

$$\frac{dy}{dx} = 0.6112x^1 - 2.7544x^0 + 0x^{-1}$$

$$\frac{dy}{dx} = 0.6112x - 2.7544$$

At the point at which the curve is at its minimum point, the gradient is equal to zero, and this equation can be equal to zero, and solved for x.

$$0.6112x - 2.7544 = 0 \qquad\qquad 0.6112x = +2.7544$$

$$x = \frac{2.7544}{0.6112} = 4.5065$$

The pH value for which the protein concentration is at its lowest is 4.5. To find the minimum protein concentration, this value can be put into the original equation:

$$y = 0.3056x^2 - 2.7544x + 6.3372$$

$$y = 0.3056 \times (4.5065)^2 - (2.7544 \times 4.5065) + 6.3372$$

$$y = (0.3056 \times 20.3085) - 12.4127 + 6.3372$$

$$y = 6.2062 - 12.4127 + 6.3372$$

$$y = 0.1308$$

While you would give the answer to finding the pH as 4.5, you would use 4.5065 to put back into the original equation to find the minimum protein concentration, thereby retaining as much accuracy as you can throughout the calculation.

The minimum protein concentration is 0.13 mg ml^{-1}.

While you can estimate these values, this method uses mathematics to provide more accurate values. You must note, however, that the values are only as accurate as the data you originally collected, and the significant figures you quote must reflect the accuracy of the original data. While you can complete the calculation with as many decimal places as possible, it is the final answer where you must think about the number of significant figures you provide.

Another example:

When a fawn is born it loses weight during the first few days and then its weight starts to increase again. The following graph shows how the mass of a fawn changed over its first 4.5 days. According to the line of the graph, what was the mass of the fawn at birth and what was its minimum mass?

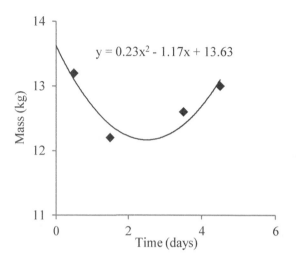

$$y = 0.23x^2 - 1.17x + 13.63$$

Fig 17.22
Mass of a fawn in its
first five days

From the equation, the intercept on the y axis shows the mass of the fawn at birth, which was 13.45 kg.

To find the minimum mass, you need to look at the <u>stationary point</u> of the line. This is where $\frac{dy}{dx} = 0$.

So you can apply the Power Rule to the equation of the line:

$$\frac{dy}{dx} = 0.46x - 1.17$$

At the stationary point $\frac{dy}{dx} = 0$, so

$$0 = 0.46x - 1.17$$

Solving for x: $1.17 = 0.46x$ $\qquad \frac{1.17}{0.46} = x$ $\qquad 2.54 = x$

So the fawn was at its lowest mass at 2.54 days. To find the mass at this time, you can substitute 2.54 for x in the original equation.

$$y = 0.23x^2 - 1.17x + 13.63$$
$$y = 0.23(2.54)^2 - (1.17 \times 2.54) + 13.63$$
$$y = 1.48 - 2.97 + 13.63$$
$$y = 12.14$$

The lowest mass of the fawn was 12.14 kg.
Always look back at the graph and check that your answer is in the right area.

The derivative of e

The derivative of e^x is a special case. This is the only function which is equal to its own derivative.

Function $\qquad\qquad y = e^x \qquad\qquad$ Derivative $\qquad\qquad \dfrac{dy}{dx} = \dfrac{de^x}{dx} = e^x$

So the derivative of $\quad y = e^{3x} \qquad$ is $\qquad\qquad \dfrac{dy}{dx} = 3e^{3x}$

and the derivative of $\quad y = 10e^{3x} \qquad$ is $\qquad\qquad \dfrac{dy}{dx} = 30e^{3x}$

Note that while the Power Rule is applied to the equation as for previous examples, with the number in front of e being multiplied by the number in front of x, the power to which e is raised does not change.

The terms x and y apply to what would be plotted on each axis of a graph, and often you will see x represented as time, or t, so that differentiating

$y = e^t \qquad\qquad\qquad$ will give you $\qquad \dfrac{dy}{dt} = e^t$

and differentiating $\quad y = e^{3t} \qquad$ will give you $\qquad \dfrac{dy}{dt} = 3e^{3t}$

You have looked at population growth with the equation $\quad N = N_0 e^{kt}$. This equation looks at how a population increases with time. Calculus can be used to determine the instantaneous increase in population size. Remember that:

$N =$ the number in the population

$t =$ time

$N_0 =$ the number in the population when $t = 0$

$k =$ growth or decay constant

$e = 2.718....$

You can differentiate the equation above using the Power Rule as it applies to equations that include e.

The letter in front of x is multiplied by the number in front of e. Everything else stays the same:

$$\dfrac{dN}{dt} = kN_0 e^{kt}$$

Now you already know that $N = N_0 e^{kt}$, and in the above equation the expression $N_0 e^{kt}$ appears in front of k, and this expression is equal to N. Because of this, N can be substituted for $N_0 e^{kt}$ in the above equation, and this gives you:

$$\frac{dN}{dt} = kN$$

Here the instantaneous rate of growth of a population is dependent on the growth rate and the number in the population.

This equation can be applied to specific situations, for example, the term k can be taken as the number of births, b, minus the number of deaths, d,

$$\frac{dN}{dt} = (b - d)N$$

so if there are more births than deaths, k (or b - d) would be positive and the population would be increasing. If k were negative, this means that the number of deaths would exceed the number of births, and so the population would be declining. If the births and deaths were equal, k would be equal to zero, so $\frac{dN}{dt} = 0$. This means that the tangent to the growth curve at this point would be horizontal.

In the above examples the letter in front of N is k. However, there are many uses of this equation and you may see a different letter used in place of k. For example, you might come across this equation written as $\frac{dN}{dt} = \mu N$ where μ is referred to as the growth rate of a population. Another example is $\frac{dN}{dt} = rN$ where r is referred to as the intrinsic rate of population growth.

This equation can form the base of a model onto which other components are added, to more accurately represent what is happening to a particular population.

For example:

$$\frac{dN}{dt} = rN \left(\frac{K - N}{K}\right)$$

where K refers to the carrying capacity of a population, or the average number of individuals an environment can support. The graph for this equation is an S-shaped curve. Changes that are made to the basic equation reflect different ways in which a population grows and will therefore result in different shaped curves on a graph.

Test your understanding

1. Determine the equation of a line which passes through the points A, B with the following coordinates:

a.   A (1, 3)   B (3 ,8)

b.   A (-4, -2)   B (-2 3)

c.   A (2, 4)   B (3, 1)

d.   A (3, 5)   B (5, 5)

e.   A (2.4, 4.5)   B (4.8, 5.7)

f.   A (-3.6, 7.9)   B (4.1, 5.3)

g.   A (0.51, 3.48) B (0.67, 6.39)

2. Differentiate the following with respect to x:

a. $y = x^2$

b. $y = -4x^3$

c. $y = 3x^2$

d. $y = x^2 + 4$

e. $y = 3x^2 - 7$

f. $y = -2x^2 + 3$

g. $y = 2x^2 + 4x - 3$

h. $y = -2x^3 + 3x + 9$

i. $y = 3x^2 - 4x - 2$

j. $y = x^{-1}$

k. $y = 2x^{-2}$

l. $y = -2x^{-1}$

m. $y = e^{2x}$

n. $y = 2e^x$

o. $y = 3e^{4x}$

3. Differentiate the following with respect to t:

a. $y = t^2 + 5$

b. $y = 2t^2 - 2$

c. $y = -2t^2 + 8$

d. $y = 2t^2 - 7t + 3$

e. $y = -3t^3 - 4t + 1$

f. $y = 2t^4 + 7t - 3$

g. $y = t^{-2}$

h. $y = 2t^{-1}$

i. $y = 2e^{3t}$

4. Plot the following data opposite in a spread sheet software package. It gives the growth of rhizoids of *Chara* at different copper concentrations. Determine the concentration of copper that provides the maximum growth of this plant, and the maximum growth at this copper concentration.

Copper concentration (µg l⁻¹)	Rhizoid growth (µm d⁻¹)
0	0
10	10
20	50
30	80
40	80
50	100
60	100
70	80
80	60
90	60
100	0

5. In the disease Malaria, protist parasites, referred to as merozoites, are released into the blood stream where they have been dividing in red blood cells. The merozoites then re enter red blood cells. So the number of merozoites in the blood stream increase to a maximum and then decrease again in a cyclical pattern. The following graph represents one cycle:

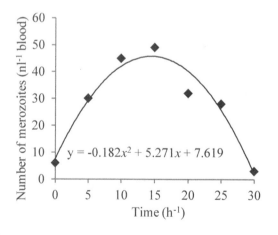

Fig 17.23
Merozoite concentration in the blood over time

Using the equation of the line, calculate:

a) the time at which the maximum number of merozoites were found in the blood.

b) what this maximum concentration was.

The graph shows: Number of merozoites (nl⁻¹ blood) on the y-axis (0 to 60) and Time (h⁻¹) on the x-axis (0 to 30). Equation: $y = -0.182x^2 + 5.271x + 7.619$

6. In the equation $\frac{dN}{dt} = (b - d)N$ where b represents birth rate and d represents death rate, and N is the number in a population and t is time, are the following statements true or false?

a. As d decreases, $\frac{dN}{dt}$ increases.

b. As b decreases, $\frac{dN}{dt}$ increases.

Solutions

Chapter 1

Self-assessment questions 1.1

1. An integer is a whole number, that is a number from the set

$$... -4, -3, -2, -1, 0, 1, 2, 3, 4 ...$$

The positive integers are 1, 2, 3, 4 .... The negative integers are ... − 4, −3, −2, −1.

2. The sum is the result of adding numbers. The difference is found by subtracting one number from another. The product of numbers is found by multiplying the numbers. The quotient of two numbers is found by dividing one number by the other.

3. (a) positive (b) negative (c) negative (d) positive

Exercise 1.1

1. (a) 3 (b) 9 (c) 11 (d) 21 (e) 30 (f) 56 (g) −13 (h) −19 (i) −19 (j) −13 (k) 29 (l) −75 (m) −75 (n) 29

2. (a) −24 (b) −32 (c) −30 (d) 16 (e) −42

3. (a) −5 (b) 3 (c) −3 (d) 3 (e) −3 (f) −6 (g) 6 (h) −6

4. (a) Sum $= 3 + 6 = 9$
   Product $= 3 \times 6 = 18$

(b) Sum $= 17$, Product $= 70$

(c) Sum $= 2 + 3 + 6 = 11$
   Product $= 2 \times 3 \times 6 = 36$

5. (a) Difference $= 18 - 9 = 9$
   Quotient $= \frac{18}{9} = 2$

(b) Difference $= 20 - 5 = 15$
   Quotient $= \frac{20}{5} = 4$

(c) Difference $= 100 - 20 = 80$
   Quotient $= \frac{100}{20} = 5$

Self-assessment questions 1.2

1. BODMAS is a priority rule used when evaluating expressions: Brackets (do first), Of, Division, Multiplication (do secondly), Addition, Subtraction (do thirdly).

2. False. For example, $(12 - 4) - 3$ is not the same as $12 - (4 - 3)$. The former is equal to 5 whereas the latter is equal to 11. The position of the brackets is clearly important.

Exercise 1.2

1. (a) $6 - 2 \times 2 = 6 - 4 = 2$
   (b) $(6 - 2) \times 2 = 4 \times 2 = 8$
   (c) $6 \div 2 - 2 = 3 - 2 = 1$
   (d) $(6 \div 2) - 2 = 3 - 2 = 1$
   (e) $6 - 2 + 3 \times 2 = 6 - 2 + 6 = 10$
   (f) $6 - (2 + 3) \times 2 = 6 - 5 \times 2$
       $= 6 - 10 = -4$

(g) $(6 - 2) + 3 \times 2 = 4 + 3 \times 2 = 4 + 6$
$= 10$

(h) $\frac{16}{-2} = -8$ (i) $\frac{-24}{-3} = 8$

(j) $(-6) \times (-2) = 12$

(k) $(-2)(-3)(-4) = -24$

2. (a) $6 \times (12 - 3) + 1 = 6 \times 9 + 1$
$= 54 + 1 = 55$

(b) $6 \times 12 - (3 + 1) = 72 - 4 = 68$

(c) $6 \times (12 - 3 + 1) = 6 \times 10 = 60$

(d) $5 \times (4 - 3) + 2 = 5 \times 1 + 2 = 5 + 2 = 7$

(e) $5 \times 4 - (3 + 2) = 20 - 5 = 15$
or $5 \times (4 - 3 + 2) = 5 \times 3 = 15$

(f) $5 \times (4 - (3 + 2)) = 5 \times (4 - 5)$
$= 5 \times (-1) = -5$

Self-assessment questions 1.3

1. A prime number is a positive integer larger than 1 that cannot be expressed as the product of two smaller positive integers.

2. 2, 3, 5, 7, 11, 13, 17, 19, 23, 29

3. All even numbers have 2 as a factor and so can be expressed as the product of two smaller numbers. The exception to this is 2 itself, which can only be expressed as 1×2, and since these numbers are not both smaller than 2, then 2 is prime.

Exercise 1.3

1. 13, 2 and 29 are prime.

2. (a) 2×13 (b) $2 \times 2 \times 5 \times 5$
(c) $3 \times 3 \times 3$ (d) 71
(e) $2 \times 2 \times 2 \times 2 \times 2 \times 2$ (f) 3×29
(g) 19×23 (h) 29×31

3. $30 = 2 \times 3 \times 5$
$42 = 2 \times 3 \times 7$
2 and 3 are common prime factors.

Self-assessment questions 1.4

1. H.c.f. stands for 'highest common factor'. The h.c.f. is the largest number that is a factor of each of the numbers in the original set.

2. L.c.m. stands for 'lowest common multiple'. It is the smallest number that can be divided exactly by each of the numbers in the set.

Exercise 1.4

1. (a) $12 = 2 \times 2 \times 3$ $15 = 3 \times 5$
$21 = 3 \times 7$
Hence h.c.f. $= 3$

(b) $16 = 2 \times 2 \times 2 \times 2$
$24 = 2 \times 2 \times 2 \times 3$
$40 = 2 \times 2 \times 2 \times 5$
So h.c.f. $= 2 \times 2 \times 2 = 8$

(c) $28 = 2 \times 2 \times 7$ $70 = 2 \times 5 \times 7$
$120 = 2 \times 2 \times 2 \times 3 \times 5$
$160 = 2 \times 2 \times 2 \times 2 \times 2 \times 5$
So h.c.f. $= 2$

(d) $35 = 5 \times 7$ $38 = 2 \times 19$
$42 = 2 \times 3 \times 7$
So h.c.f. $= 1$

(e) $96 = 2 \times 2 \times 2 \times 2 \times 2 \times 3$
$120 = 2 \times 2 \times 2 \times 3 \times 5$
$144 = 2 \times 2 \times 2 \times 2 \times 3 \times 3$
So h.c.f. $= 2 \times 2 \times 2 \times 3 = 24$

2. (a) 5 $6 = 2 \times 3$ $8 = 2 \times 2 \times 2$
So l.c.m. $= 2 \times 2 \times 2 \times 3 \times 5$
$= 120$

(b) $20 = 2 \times 2 \times 5$ $30 = 2 \times 3 \times 5$
So l.c.m. $= 2 \times 2 \times 3 \times 5 = 60$

(c) 7 $9 = 3 \times 3$ $12 = 2 \times 2 \times 3$
So l.c.m. $= 2 \times 2 \times 3 \times 3 \times 7$
$= 252$

(d) $100 = 2 \times 2 \times 5 \times 5$
$150 = 2 \times 3 \times 5 \times 5$
$235 = 5 \times 47$
So l.c.m. $=$
$2 \times 2 \times 3 \times 5 \times 5 \times 47 = 14100$

(e) $96 = 2 \times 2 \times 2 \times 2 \times 2 \times 3$
$120 = 2 \times 2 \times 2 \times 3 \times 5$
$144 = 2 \times 2 \times 2 \times 2 \times 3 \times 3$
So l.c.m. $=$
$2 \times 2 \times 2 \times 2 \times 2 \times 3 \times 3 \times 5$
$= 1440$

Chapter 2

Self-assessment questions 2.1

1. (a) A fraction is formed by dividing a whole number by another whole number, for example $\frac{11}{3}$.
   (b) If the top number (the numerator) of a fraction is greater than or equal to the bottom number (the denominator) then the fraction is improper. For example, $\frac{101}{100}$ and $\frac{9}{9}$ are both improper fractions.
   (c) If the top number is less than the bottom number then the fraction is proper. For example, $\frac{99}{100}$ is a proper fraction.
2. (a) The numerator is the 'top number' of a fraction.
   (b) The denominator is the 'bottom number' of a fraction. For example, in $\frac{17}{14}$, the numerator is 17 and the denominator is 14.

Exercise 2.1

1. (a) Proper (b) Proper (c) Improper
   (d) Proper (e) Improper

Self-assessment questions 2.2

1. True. For example $7 = \frac{7}{1}$.
2. The numerator and denominator can both be divided by their h.c.f. This simplifies the fraction.
3. $\frac{3}{4}, \frac{9}{12}, \frac{75}{100}$.

Exercise 2.2

1. (a) $\frac{18}{27} = \frac{2}{3}$ (b) $\frac{12}{20} = \frac{3}{5}$ (c) $\frac{15}{45} = \frac{1}{3}$
   (d) $\frac{25}{80} = \frac{5}{16}$ (e) $\frac{15}{60} = \frac{1}{4}$ (f) $\frac{90}{200} = \frac{9}{20}$
   (g) $\frac{15}{20} = \frac{3}{4}$ (h) $\frac{2}{18} = \frac{1}{9}$ (i) $\frac{16}{24} = \frac{2}{3}$
   (j) $\frac{30}{65} = \frac{6}{13}$ (k) $\frac{12}{21} = \frac{4}{7}$ (l) $\frac{100}{45} = \frac{20}{9}$
   (m) $\frac{6}{9} = \frac{2}{3}$ (n) $\frac{12}{16} = \frac{3}{4}$ (o) $\frac{13}{42}$
   (p) $\frac{13}{39} = \frac{1}{3}$ (q) $\frac{11}{33} = \frac{1}{3}$ (r) $\frac{14}{30} = \frac{7}{15}$
   (s) $-\frac{12}{16} = -\frac{3}{4}$ (t) $\frac{11}{-33} = -\frac{1}{3}$
   (u) $\frac{-14}{-30} = \frac{7}{15}$

2. $\frac{3}{4} = \frac{21}{28}$

3. $4 = \frac{20}{5}$

4. $\frac{5}{12} = \frac{15}{36}$

5. $2 = \frac{8}{4}$

6. $6 = \frac{18}{3}$

7. $\frac{2}{3} = \frac{8}{12}, \frac{5}{4} = \frac{15}{12}, \frac{5}{6} = \frac{10}{12}$

8. $\frac{4}{9} = \frac{8}{18}, \frac{1}{2} = \frac{9}{18}, \frac{5}{6} = \frac{15}{18}$

9. (a) $\frac{1}{2} = \frac{6}{12}$ (b) $\frac{3}{4} = \frac{9}{12}$ (c) $\frac{5}{2} = \frac{30}{12}$
   (d) $5 = \frac{60}{12}$ (e) $4 = \frac{48}{12}$ (f) $12 = \frac{144}{12}$

Self-assessment question 2.3

1. The l.c.m. of the denominators is found. Each fraction is expressed in an equivalent form with this l.c.m. as denominator. Addition and subtraction can then take place.

Exercise 2.3

1. (a) $\dfrac{1}{4}+\dfrac{2}{3}=\dfrac{3}{12}+\dfrac{8}{12}=\dfrac{11}{12}$

(b) $\dfrac{3}{5}+\dfrac{5}{3}=\dfrac{9}{15}+\dfrac{25}{15}=\dfrac{34}{15}$

(c) $\dfrac{12}{14}-\dfrac{2}{7}=\dfrac{6}{7}-\dfrac{2}{7}=\dfrac{4}{7}$

(d) $\dfrac{3}{7}-\dfrac{1}{2}+\dfrac{2}{21}=\dfrac{18}{42}-\dfrac{21}{42}+\dfrac{4}{42}=\dfrac{1}{42}$

(e) $1\dfrac{1}{2}+\dfrac{4}{9}=\dfrac{3}{2}+\dfrac{4}{9}=\dfrac{27}{18}+\dfrac{8}{18}=\dfrac{35}{18}$

(f) $2\dfrac{1}{4}-1\dfrac{1}{3}+\dfrac{1}{2}=\dfrac{9}{4}-\dfrac{4}{3}+\dfrac{1}{2}$

$=\dfrac{27}{12}-\dfrac{16}{12}+\dfrac{6}{12}=\dfrac{17}{12}$

(g) $\dfrac{10}{15}-1\dfrac{2}{5}+\dfrac{8}{3}=\dfrac{10}{15}-\dfrac{7}{5}+\dfrac{8}{3}$

$=\dfrac{10}{15}-\dfrac{21}{15}+\dfrac{40}{15}=\dfrac{29}{15}$

(h) $\dfrac{9}{10}-\dfrac{7}{16}+\dfrac{1}{2}-\dfrac{2}{5}=\dfrac{72}{80}-\dfrac{35}{80}+\dfrac{40}{80}-\dfrac{32}{80}$

$=\dfrac{45}{80}=\dfrac{9}{16}$

2. (a) $\dfrac{7}{8}+\dfrac{1}{3}=\dfrac{21}{24}+\dfrac{8}{24}=\dfrac{29}{24}$

(b) $\dfrac{1}{2}-\dfrac{3}{4}=\dfrac{2}{4}-\dfrac{3}{4}=-\dfrac{1}{4}$

(c) $\dfrac{3}{5}+\dfrac{2}{3}+\dfrac{1}{2}=\dfrac{18}{30}+\dfrac{20}{30}+\dfrac{15}{30}=\dfrac{53}{30}=1\dfrac{23}{30}$

(d) $\dfrac{3}{8}+\dfrac{1}{3}+\dfrac{1}{4}=\dfrac{9}{24}+\dfrac{8}{24}+\dfrac{6}{24}=\dfrac{23}{24}$

(e) $\dfrac{2}{3}-\dfrac{4}{7}=\dfrac{14}{21}-\dfrac{12}{21}=\dfrac{2}{21}$

(f) $\dfrac{1}{11}-\dfrac{1}{2}=\dfrac{2}{22}-\dfrac{11}{22}=-\dfrac{9}{22}$

(g) $\dfrac{3}{11}-\dfrac{5}{8}=\dfrac{24}{88}-\dfrac{55}{88}=\dfrac{-31}{88}$

3. (a) $\dfrac{5}{2}$ (b) $\dfrac{11}{3}$ (c) $\dfrac{41}{4}$ (d) $\dfrac{37}{7}$ (e) $\dfrac{56}{9}$

(f) $\dfrac{34}{3}$ (g) $\dfrac{31}{2}$ (h) $\dfrac{55}{4}$ (i) $\dfrac{133}{11}$ (j) $\dfrac{41}{3}$

(k) $\dfrac{113}{2}$

4. (a) $3\dfrac{1}{3}$ (b) $3\dfrac{1}{2}$ (c) $3\dfrac{3}{4}$ (d) $4\dfrac{1}{6}$

Self-assessment question 2.4

1. The numerators are multiplied together to form the numerator of the product. The denominators are multiplied to form the denominator of the product.

Exercise 2.4

1. (a) $\dfrac{2}{3}\times\dfrac{6}{7}=\dfrac{2}{1}\times\dfrac{2}{7}=\dfrac{4}{7}$

(b) $\dfrac{8}{15}\times\dfrac{25}{32}=\dfrac{1}{15}\times\dfrac{25}{4}=\dfrac{1}{3}\times\dfrac{5}{4}=\dfrac{5}{12}$

(c) $\dfrac{1}{4}\times\dfrac{8}{9}=\dfrac{1}{1}\times\dfrac{2}{9}=\dfrac{2}{9}$

(d) $\dfrac{16}{17}\times\dfrac{34}{48}=\dfrac{1}{17}\times\dfrac{34}{3}=\dfrac{1}{1}\times\dfrac{2}{3}=\dfrac{2}{3}$

(e) $2\times\dfrac{3}{5}\times\dfrac{5}{12}=\dfrac{3}{5}\times\dfrac{5}{6}=\dfrac{3}{1}\times\dfrac{1}{6}=\dfrac{1}{2}$

(f) $2\dfrac{1}{3}\times1\dfrac{1}{4}=\dfrac{7}{3}\times\dfrac{5}{4}=\dfrac{35}{12}$

(g) $1\dfrac{3}{4}\times2\dfrac{1}{2}=\dfrac{7}{4}\times\dfrac{5}{2}=\dfrac{35}{8}$

(h) $\dfrac{3}{4}\times1\dfrac{1}{2}\times3\dfrac{1}{2}=\dfrac{3}{4}\times\dfrac{3}{2}\times\dfrac{7}{2}=\dfrac{63}{16}$

2. (a) $\dfrac{2}{3}\times\dfrac{3}{4}=\dfrac{1}{2}$ (b) $\dfrac{4}{7}\times\dfrac{21}{30}=\dfrac{6}{15}=\dfrac{2}{5}$

(c) $\dfrac{9}{10}\times80=72$ (d) $\dfrac{6}{7}\times42=36$

3. Yes, because $\dfrac{3}{4} \times \dfrac{12}{15} = \dfrac{12}{15} \times \dfrac{3}{4}$

4. (a) $-\dfrac{5}{21}$ (b) $-\dfrac{3}{8}$ (c) $-\dfrac{5}{11}$ (d) $\dfrac{10}{7}$

5. (a) $5\dfrac{1}{2} \times \dfrac{1}{2} = \dfrac{11}{2} \times \dfrac{1}{2} = \dfrac{11}{4}$

(b) $3\dfrac{3}{4} \times \dfrac{1}{3} = \dfrac{15}{4} \times \dfrac{1}{3} = \dfrac{5}{4}$

(c) $\dfrac{2}{3} \times 5\dfrac{1}{9} = \dfrac{2}{3} \times \dfrac{46}{9} = \dfrac{92}{27}$

(d) $\dfrac{3}{4} \times 11\dfrac{1}{2} = \dfrac{3}{4} \times \dfrac{23}{2} = \dfrac{69}{8}$

6. (a) $\dfrac{3}{5} \times 11\dfrac{1}{4} = \dfrac{3}{5} \times \dfrac{45}{4} = \dfrac{27}{4}$

(b) $\dfrac{2}{3} \times 15\dfrac{1}{2} = \dfrac{2}{3} \times \dfrac{31}{2} = \dfrac{31}{3}$

(c) $\dfrac{1}{4} \times \left(-8\dfrac{1}{3}\right) = \dfrac{1}{4} \times \left(-\dfrac{25}{3}\right) = -\dfrac{25}{12}$

Self-assessment question 2.5

1. To divide one fraction by a second fraction, the second fraction is inverted (that is, the numerator and denominator are interchanged) and then multiplication is performed.

Exercise 2.5

1. (a) $\dfrac{3}{4} \div \dfrac{1}{8} = \dfrac{3}{4} \times \dfrac{8}{1} = \dfrac{3}{1} \times \dfrac{2}{1} = 6$

(b) $\dfrac{8}{9} \div \dfrac{4}{3} = \dfrac{8}{9} \times \dfrac{3}{4} = \dfrac{2}{9} \times \dfrac{3}{1} = \dfrac{2}{3}$

(c) $-\dfrac{2}{7} \div \dfrac{4}{21} = -\dfrac{2}{7} \times \dfrac{21}{4} = -\dfrac{2}{1} \times \dfrac{3}{4} = -\dfrac{3}{2}$

(d) $\dfrac{9}{4} \div 1\dfrac{1}{2} = \dfrac{9}{4} \div \dfrac{3}{2} = \dfrac{9}{4} \times \dfrac{2}{3} = \dfrac{3}{4} \times \dfrac{2}{1} = \dfrac{3}{2}$

(e) $\dfrac{5}{6} \div \dfrac{5}{12} = \dfrac{5}{6} \times \dfrac{12}{5} = \dfrac{1}{6} \times \dfrac{12}{1} = 2$

(f) $\dfrac{99}{100} \div 1\dfrac{4}{5} = \dfrac{99}{100} \div \dfrac{9}{5} = \dfrac{99}{100} \times \dfrac{5}{9}$
$= \dfrac{11}{100} \times \dfrac{5}{1} = \dfrac{11}{20}$

(g) $3\dfrac{1}{4} \div 1\dfrac{1}{8} = \dfrac{13}{4} \div \dfrac{9}{8} = \dfrac{13}{4} \times \dfrac{8}{9}$
$= \dfrac{13}{1} \times \dfrac{2}{9} = \dfrac{26}{9}$

(h) $\left(2\dfrac{1}{4} \div \dfrac{3}{4}\right) \times 2 = \left(\dfrac{9}{4} \times \dfrac{4}{3}\right) \times 2$
$= \left(\dfrac{3}{4} \times 4\right) \times 2 = 3 \times 2 = 6$

(i) $2\dfrac{1}{4} \div \left(\dfrac{3}{4} \times 2\right) = \dfrac{9}{4} \div \left(\dfrac{3}{4} \times \dfrac{2}{1}\right)$
$= \dfrac{9}{4} \div \dfrac{3}{2} = \dfrac{9}{4} \times \dfrac{2}{3} = \dfrac{3}{4} \times \dfrac{2}{1} = \dfrac{3}{2}$

(j) $6\dfrac{1}{4} \div 2\dfrac{1}{2} + 5 = \dfrac{25}{4} \div \dfrac{5}{2} + 5$
$= \dfrac{25}{4} \times \dfrac{2}{5} + 5 = \dfrac{5}{2} + 5 = \dfrac{15}{2}$

(k) $6\dfrac{1}{4} \div \left(2\dfrac{1}{2} + 5\right) = \dfrac{25}{4} \div \left(\dfrac{5}{2} + 5\right)$
$= \dfrac{25}{4} \div \dfrac{15}{2} = \dfrac{25}{4} \times \dfrac{2}{15} = \dfrac{5}{4} \times \dfrac{2}{3} = \dfrac{5}{6}$

Chapter 3

Self-assessment questions 3.1

1. Largest is 23.01; smallest is 23.0.
2. 0.1

Exercise 3.1

1. (a) $\dfrac{7}{10}$ (b) $\dfrac{4}{5}$ (c) $\dfrac{9}{10}$

2. (a) $\dfrac{11}{20}$ (b) $\dfrac{79}{500}$ (c) $\dfrac{49}{50}$ (d) $\dfrac{99}{1000}$

3. (a) $4\dfrac{3}{5}$ (b) $5\dfrac{1}{5}$ (c) $8\dfrac{1}{20}$ (d) $11\dfrac{59}{100}$

   (e) $121\dfrac{9}{100}$

4. (a) 0.697 (b) 0.083 (c) 0.517

Self-assessment questions 3.2

1. Writing a number to 2 (or 3 or 4 etc.) significant figures is a way of approximating

the number. The number of s.f. is the maximum number of non-zero digits in the approximation. The approximation is as close as possible to the original number.
2. The digits after the decimal point are considered. To write to 1 d.p. we consider the first two digits; to write to 2 d.p. we consider the first three digits, and so on. If the last digit considered is 5 or greater, we round up the previous digit; otherwise we round down.

Exercise 3.2

1. (a) 6960 (b) 70.4 (c) 0.0123 (d) 0.0110 (e) 45.6 (f) 2350
2. (a) 66.00 (b) 66.0 (c) 66 (d) 70 (e) 66.00 (f) 66.0
3. (a) 10 (b) 10.0
4. (a) 65.456 (b) 65.46 (c) 65.5 (d) 65.456 (e) 65.46 (f) 65.5 (g) 65 (h) 70

Chapter 4

Self-assessment question 4.1

1. Converting fractions to percentages allows for easy comparison of numbers.

Exercise 4.1

1. 23% of 124 $= \dfrac{23}{100} \times 124 = 28.52$

2. (a) $\dfrac{9}{11} = \dfrac{9}{11} \times 100\% = \dfrac{900}{11}\% = 81.82\%$

   (b) $\dfrac{15}{20} = \dfrac{15}{20} \times 100\% = 75\%$

   (c) $\dfrac{9}{10} = \dfrac{9}{10} \times 100\% = 90\%$

   (d) $\dfrac{45}{50} = \dfrac{45}{50} \times 100\% = 90\%$

   (e) $\dfrac{75}{90} = \dfrac{75}{90} \times 100\% = 83.33\%$

3. $\dfrac{13}{12} = \dfrac{13}{12} \times 100\% = 108.33\%$

4. 217% of 500 $= \dfrac{217}{100} \times 500 = 1085$

5. New weekly wage is 106% of £400,

   106% of 400 $= \dfrac{106}{100} \times 400 = 424$

   The new weekly wage is £424.

6. 17% of $1200 = \dfrac{17}{100} \times 1200 = 204$

   The debt is decreased by £204 to
   £1200 − 204 = £996.

7. (a) $50\% = \dfrac{50}{100} = 0.5$

   (b) $36\% = \dfrac{36}{100} = 0.36$

   (c) $75\% = \dfrac{75}{100} = 0.75$

   (d) $100\% = \dfrac{100}{100} = 1$

   (e) $12.5\% = \dfrac{12.5}{100} = 0.125$

8. £204.80

9. £1125

10. percentage change

$$= \dfrac{\text{new value} - \text{original value}}{\text{original value}} \times 100$$

$$= \dfrac{7495 - 6950}{6950} \times 100$$

$$= 7.84\%$$

11. percentage change

$$= \dfrac{\text{new value} - \text{original value}}{\text{original value}} \times 100$$

$$= \dfrac{399 - 525}{525} \times 100$$

$$= -24\%$$

Note that the percentage change is negative and this indicates a reduction in price. There has been a 24% reduction in the price of the washing machine.

Self-assessment question 4.2

1. True

Exercise 4.2

1. $8 + 1 + 3 = 12$. The first number is $\frac{8}{12}$ of 180, that is 120; the second number is $\frac{1}{12}$ of 180, that is 15; and the third number is $\frac{3}{12}$ of 180, that is 45. Hence 180 is divided into 120, 15 and 45.

2. $1 + 1 + 3 = 5$. We calculate $\frac{1}{5}$ of 930 to be 186 and $\frac{3}{5}$ of 930 to be 558. The length is divided into 186 cm, 186 cm and 558 cm.

3. $2 + 3 + 4 = 9$. The first piece is $\frac{2}{9}$ of 6 m, that is 1.33 m; the second piece is $\frac{3}{9}$ of 6 m, that is 2 m; the third piece is $\frac{4}{9}$ of 6 m, that is 2.67 m.

4. $1 + 2 + 3 + 4 = 10$: $\frac{1}{10}$ of 1200 = 120; $\frac{2}{10}$ of 1200 = 240; $\frac{3}{10}$ of 1200 = 360; $\frac{4}{10}$ of 1200 = 480. The number 1200 is divided into 120, 240, 360 and 480.

5. $2\frac{3}{4} : 1\frac{1}{2} : 2\frac{1}{4} = \dfrac{11}{4} : \dfrac{3}{2} : \dfrac{9}{4} = 11 : 6 : 9$

   Now, $11 + 6 + 9 = 26$, so

$$\dfrac{11}{26} \times 2600 = 1100 \qquad \dfrac{6}{26} \times 2600 = 600$$

$$\dfrac{9}{26} \times 2600 = 900$$

   Alan receives £1100, Bill receives £600 and Claire receives £900.

6. 8 kg, 10.67 kg, 21.33 kg

7. (a) $1 : 2$ (b) $1 : 2$ (c) $1 : 2 : 4$ (d) $1 : 21$

8. 24, 84

Chapter 5

Self-assessment questions 5.1

1. 'Algebra' refers to the manipulation of symbols, as opposed to the manipulation of numbers.

2. The product of the two numbers is written as *mn*.

3. An algebraic fraction is formed by dividing one algebraic expression by another algebraic expression. The 'top' of the fraction is the numerator; the 'bottom' of the fraction is the denominator.

4. Superscripts and subscripts are located in different positions, relative to the symbol. Superscripts are placed high; subscripts are placed low.
5. A variable can have many different values; a constant has one, fixed value.

Self-assessment questions 5.2

1. In the expression a^x, a is the base and x is the power.
2. 'Index' is another word meaning 'power'.
3. $(xyz)^2$ means $(xyz)(xyz)$, which can be written as $x^2y^2z^2$. Clearly this is distinct from xyz^2, in which only the quantity z is squared.
4. $(-3)^4 = (-3)(-3)(-3)(-3) = 81$.
   $-3^4 = -(3)(3)(3)(3) = -81$. Here, the power (4) has the higher priority.

Exercise 5.2

1. $2^4 = 16$; $(\frac{1}{2})^2 = \frac{1}{4}$; $1^8 = 1$; $3^5 = 243$;
   $0^3 = 0$.
2. $10^4 = 10000$; $10^5 = 100000$;
   $10^6 = 1000000$.
3. $11^4 = 14641$; $16^8 = 4294967296$;
   $39^4 = 2313441$; $1.5^7 = 17.0859375$.
4. (a) $a^4b^2c = a \times a \times a \times a \times b \times b \times c$
   (b) $xy^2z^4 = x \times y \times y \times z \times z \times z \times z$
5. (a) x^4y^2 (b) $x^2y^2z^3$ (c) $x^2y^2z^2$
   (d) $a^2b^2c^2$
6. (a) $7^4 = 2401$ (b) $7^5 = 16807$
   (c) $7^4 \times 7^5 = 40353607$
   (d) $7^9 = 40353607$ (e) $8^3 = 512$
   (f) $8^7 = 2097152$
   (g) $8^3 \times 8^7 = 1073741824$
   (h) $8^{10} = 1073741824$
   The rule states that $a^m \times a^n = a^{m+n}$; the powers are added.
7. $(-3)^3 = -27$; $(-2)^2 = 4$; $(-1)^7 = -1$;
   $(-1)^4 = 1$.
8. -4492.125; 324; -0.03125.
9. (a) 36   (b) 9   (c) -64 (d) -8
   $-6^2 = -36$, $-3^2 = -9$,
   $-4^3 = -64$, $-2^3 = -8$

Self-assessment question 5.3

1. An algebraic expression is any quantity comprising symbols and operations (that is, $+$, $-$, $\times$, $\div$): for example, x, $2x^2$ and $3x + 2y$ are all algebraic expressions. An algebraic formula relates two or more quantities and must contain an '$=$' sign. For example, $A = \pi r^2$, $V = \pi r^2 h$ and $S = ut + \frac{1}{2}at^2$ are all algebraic formulae.

Exercise 5.3

1. 60
2. 69
3. (a) 314.2 cm^2 (b) 28.28 cm^2
   (c) 0.126 cm^2
4. $3x^2 = 3 \times 4^2 = 3 \times 16 = 48$;
   $(3x)^2 = 12^2 = 144$
5. $5x^2 = 5(-2)^2 = 20$; $(5x)^2 = (-10)^2 = 100$
6. (a) 33.95   (b) 23.5225   (c) 26.75
   (d) 109.234125
7. (a) $a + b + c = 25.5$ (b) $ab = 46.08$
   (c) $bc = 32.76$ (d) $abc = 419.328$
8. $C = \frac{5}{9}(100 - 32) = \frac{5}{9}(68) = 37.78$
9. (a) $x^2 = 49$ (b) $-x^2 = -49$
   (c) $(-x)^2 = (-7)^2 = 49$
10. (a) 4   (b) 4   (c) -4 (d) 12
    (e) -12 (f) 36
11. (a) 3   (b) 9   (c) -1 (d) 36
    (e) -36 (f) 144
12. $x^2 - 7x + 2 = (-9)^2 - 7(-9) + 2$
    $= 81 + 63 + 2 = 146$
13. $2x^2 + 3x - 11 = 2(-3)^2 + 3(-3) - 11$
    $= 18 - 9 - 11 = -2$
14. $-x^2 + 3x - 5 = -(-1)^2 + 3(-1) - 5$
    $= -1 - 3 - 5 = -9$
15. 0
16. (a) 49   (b) 43   (c) 1
    (d) 5
17. (a) 21   (b) 27   (c) 0
    (d) $\frac{1}{6}$
18. (a) 3   (b) $3\frac{4}{5}$ (c) 23   (d) 23

19. (a) $-\dfrac{1}{2}$ (b) 4 (c) 32

20. (a) 0 (b) 16 (c) $40\dfrac{1}{2}$

    (d) $2\dfrac{1}{2}$

21. (a) 17 (b) -0.5 (c) 7
22. (a) 6000 (b) 2812.5
23. (a) 17151 (b) 276951

Chapter 6

Self-assessment questions 6.1

1. $a^m \times a^n = a^{m+n}$
   $\dfrac{a^m}{a^n} = a^{m-n}$
   $(a^m)^n = a^{mn}$
2. a^0 is 1.
3. x^1 is simply x.

Exercise 6.1

1. (a) $5^7 \times 5^{13} = 5^{20}$ (b) $9^8 \times 9^5 = 9^{13}$
   (c) $11^2 \times 11^3 \times 11^4 = 11^9$
2. (a) $15^3/15^2 = 15^1 = 15$ (b) $4^{18}/4^9 = 4^9$
   (c) $5^{20}/5^{19} = 5^1 = 5$
3. (a) a^{10} (b) a^9 (c) b^{22}
4. (a) x^{15} (b) y^{21}
5. $19^8 \times 17^8$ cannot be simplified using the laws of indices because the two bases are not the same.
6. (a) $(7^3)^2 = 7^{3\times2} = 7^6$ (b) $(4^2)^8 = 4^{16}$
   (c) $(7^9)^2 = 7^{18}$
7. $1/(5^3)^8 = 1/5^{24}$
8. (a) x^5y^5 (b) $a^3b^3c^3$
9. (a) $x^{10}y^{20}$ (b) $81x^6$ (c) $-27x^3$
   (d) x^8y^{12}
10. (a) z^3 (b) y^2 (c) 1

Self-assessment question 6.2

1. a^{-m} is the same as $\dfrac{1}{a^m}$. For example, 5^{-2} is the same as $\dfrac{1}{5^2}$.

Exercise 6.2

1. (a) $\dfrac{1}{4}$ (b) $\dfrac{1}{8}$ (c) $\dfrac{1}{9}$ (d) $\dfrac{1}{27}$ (e) $\dfrac{1}{25}$
   (f) $\dfrac{1}{16}$ (g) $\dfrac{1}{9}$ (h) $\dfrac{1}{121}$ (i) $\dfrac{1}{7}$
2. (a) 0.1 (b) 0.01 (c) 0.000001 (d) 0.01
   (e) 0.001 (f) 0.0001
3. (a) $\dfrac{1}{x^4}$ (b) x^5 (c) $\dfrac{1}{x^7}$ (d) $\dfrac{1}{y^2}$
   (e) $y^1 = y$ (f) $\dfrac{1}{y^1} = \dfrac{1}{y}$ (g) $\dfrac{1}{y^2}$
   (h) $\dfrac{1}{z^1} = \dfrac{1}{z}$ (i) $z^1 = z$
4. (a) $x^{-3} = \dfrac{1}{x^3}$ (b) $x^{-5} = \dfrac{1}{x^5}$ (c) $x^{-1} = \dfrac{1}{x}$
   (d) x^5 (e) $x^{-13} = \dfrac{1}{x^{13}}$ (f) $x^{-8} = \dfrac{1}{x^8}$
   (g) $x^{-9} = \dfrac{1}{x^9}$ (h) $x^{-4} = \dfrac{1}{x^4}$
5. (a) a^{11} (b) x^{-16} (c) x^{-18} (d) 4^{-6}
6. (a) 0.001 (b) 0.0001 (c) 0.00001
7. $4^{-8}/4^{-6} = 4^{-2} = 1/4^2 = \frac{1}{16}$
   $3^{-5}/3^{-8} = 3^3 = 27$

Self-assessment questions 6.3

1. $x^{\frac{1}{2}}$ is the square root of x. $x^{\frac{1}{3}}$ is the cube root of x.

2. $10, -10$. Negative numbers, such as -100, do not have square roots, since squaring a number always gives a positive result.

Exercise 6.3

1. (a) $64^{1/3} = \sqrt[3]{64} = 4$ since $4^3 = 64$
   (b) $144^{1/2} = \sqrt{144} = \pm 12$
   (c) $16^{-1/4} = 1/16^{1/4} = 1/\sqrt[4]{16} = \pm\frac{1}{2}$
   (d) $25^{-1/2} = 1/25^{1/2} = 1/\sqrt{25} = \pm\frac{1}{5}$
   (e) $1/32^{-1/5} = 32^{1/5} = \sqrt[5]{32} = 2$ since $2^5 = 32$
2. (a) $(3^{-1/2})^4 = 3^{-2} = 1/3^2 = \frac{1}{9}$
   (b) $(8^{1/3})^{-1} = 8^{-1/3} = 1/8^{1/3} = 1/\sqrt[3]{8} = \frac{1}{2}$
3. (a) $8^{1/2}$ (b) $12^{1/3}$
   (c) $16^{1/4}$ (d) $13^{3/2}$ (e) $4^{7/3}$
4. (a) $x^{1/2}$ (b) $y^{1/3}$
   (c) $x^{5/2}$ (d) $5^{7/3}$

Exercise 6.4

1. (a) 743
   (b) 74300
   (c) 70
   (d) 0.0007
2. (a) 3×10^2
   (b) 3.56×10^2
   (c) 0.32×10^2
   (d) 0.0057×10^2

Self-assessment question 6.5

1. Scientific notation is useful for writing very large or extremely small numbers in a concise way. It is easier to manipulate such numbers when they are written using scientific notation.

Exercise 6.5

1. (a) $45 = 4.5 \times 10^1$
   (b) $45000 = 4.5 \times 10^4$
   (c) $-450 = -4.5 \times 10^2$
   (d) $90000000 = 9.0 \times 10^7$
   (e) $0.15 = 1.5 \times 10^{-1}$
   (f) $0.00036 = 3.6 \times 10^{-4}$
   (g) 3.5 is already in standard form.
   (h) $-13.2 = -1.32 \times 10^1$
   (i) $1000000 = 1 \times 10^6$
   (j) $0.0975 = 9.75 \times 10^{-2}$
   (k) $45.34 = 4.534 \times 10^1$
2. (a) $3.75 \times 10^2 = 375$
   (b) $3.97 \times 10^1 = 39.7$
   (c) $1.875 \times 10^{-1} = 0.1875$
   (d) $-8.75 \times 10^{-3} = -0.00875$
3. (a) 2.4×10^8 (b) 7.968×10^8
   (c) 1.044×10^{-4}
   (d) 1.526×10^{-1}
   (e) 5.293×10^2

Chapter 7

Exercise 7.1

1. (a) $-5p + 19q$ (b) $-5r - 13s + z$ (c) not possible to simplify
   (d) $8x^2 + 3y^2 - 2y$ (e) $4x^2 - x + 9$
2. (a) $-12y + 8p + 9q$ (b) $21x^2 - 11x^3 + y^3$
   (c) $7xy + y^2$ (d) $2xy$ (e) 0

Self-assessment questions 7.2

1. Positive
2. Negative

Exercise 7.2

1. (a) 84 (b) 84 (c) 84
2. (a) 40 (b) 40
3. (a) $14z$ (b) $30y$ (c) $6x$ (d) $27a$
   (e) $55a$ (f) $6x$
4. (a) $20x^2$ (b) $6y^3$ (c) $22u^2$ (d) $8u^2$
   (e) $26z^2$
5. (a) $21x^2$ (b) $21a^2$ (c) $14a^2$
6. (a) $15y^2$ (b) $8y$
7. (a) $a^3b^2c^2$ (b) x^3y^2 (c) x^2y^4
8. No difference; both equal x^2y^4.

9. $(xy^2)(xy^2) = x^2y^4$; $xy^2 + xy^2 = 2xy^2$
10. (a) $-21z^2$ (b) $-4z$
11. (a) $-3x^2$ (b) $2x$
12. (a) $2x^2$ (b) $-3x$

Exercise 7.3

1. (a) $4x + 4$ (b) $-4x - 4$ (c) $4x - 4$
   (d) $-4x + 4$
2. (a) $5x - 5y$ (b) $19x + 57y$ (c) $8a + 8b$
   (d) $5y + xy$ (e) $12x + 48$
   (f) $17x - 153$ (g) $-a + 2b$ (h) $x + \frac{1}{2}$
   (i) $6m - 12m^2 - 9mn$
3. (a) $18 - 13x - 26 = -8 - 13x$ (b) $x^2 + xy$
   will not simplify any further

4. (a) $x^2 + 7x + 6$ (b) $x^2 + 9x + 20$
   (c) $x^2 + x - 6$ (d) $x^2 + 5x - 6$
   (e) $xm + ym + nx + yn$
   (f) $12 + 3y + 4x + yx$ (g) $25 - x^2$
   (h) $51x^2 - 79x - 10$
5. (a) $x^2 - 4x - 21$ (b) $6x^2 + 11x - 7$
   (c) $16x^2 - 1$ (d) $x^2 - 9$
   (e) $6 + x - 2x^2$
6. (a) $\frac{29}{2}x - \frac{5}{2}y$ (b) $\frac{5}{4}x + \frac{5}{4}$
7. (a) $-x + y$ (b) $-a - 2b$ (c) $-\frac{3}{2}p - \frac{1}{2}q$
8. $(x + 1)(x + 2) = x^2 + 3x + 2.$ So
   $(x + 1)(x + 2)(x + 3) = (x^2 + 3x + 2)(x + 3)$
   $= (x^2 + 3x + 2)(x) + (x^2 + 3x + 2)(3)$
   $= x^3 + 3x^2 + 2x + 3x^2 + 9x + 6$
   $= x^3 + 6x^2 + 11x + 6$

Chapter 8

Self-assessment question 8.1

1. To factorise an expression means to write it as a product, usually of two or more simpler expressions.

Exercise 8.1

1. (a) $9x + 27$ (b) $-5x + 10$ (c) $\frac{1}{2}x + \frac{1}{2}$
   (d) $-a + 3b$ (e) $1/(2x + 2y)$
   (f) $x/(yx - y^2)$
2. (a) $4x^2$ has factors $1, 2, 4, x, 2x, 4x, x^2, 2x^2,$
       $4x^2$
   (b) $6x^3$ has factors $1, 2, 3, 6, x, 2x, 3x, 6x,$
       $x^2, 2x^2, 3x^2, 6x^2, x^3, 2x^3, 3x^3, 6x^3$
3. (a) $3(x + 6)$ (b) $3(y - 3)$ (c) $-3(y + 3)$
   (d) $-3(1 + 3y)$ (e) $5(4 + t)$ (f) $5(4 - t)$
   (g) $-5(t + 4)$ (h) $3(x + 4)$ (i) $17(t + 2)$
   (j) $4(t - 9)$
4. (a) $x(x^3 + 2)$ (b) $x(x^3 - 2)$
   (c) $x(3x^3 - 2)$ (d) $x(3x^3 + 2)$
   (e) $x^2(3x^2 + 2)$ (f) $x^3(3x + 2)$
   (g) $z(17 - z)$ (h) $x(3 - y)$ (i) $y(3 - x)$
   (j) $x(1 + 2y + 3yz)$

5. (a) $10(x + 2y)$ (b) $3(4a + b)$
   (c) $2x(2 - 3y)$ (d) $7(a + 2)$ (e) $5(2m - 3)$
   (f) $1/[5(a + 7b)]$ (g) $1/[5a(a + 7b)]$
6. (a) $3x(5x + 1)$ (b) $x(4x - 3)$
   (c) $4x(x - 2)$ (d) $3(5 - x^2)$
   (e) $5x^2(2x + 1 + 3y)$ (f) $6ab(a - 2b)$
   (g) $8b(2ac - ba + 3c)$

Self-assessment question 8.2

1. $2x^2 + x + 6$ cannot be factorised. On the other hand, $2x^2 + x - 6$ can be factorised as $(2x - 3)(x + 2)$.

Exercise 8.2

1. (a) $(x + 2)(x + 1)$ (b) $(x + 7)(x + 6)$
   (c) $(x + 5)(x - 3)$ (d) $(x + 10)(x - 1)$
   (e) $(x - 8)(x - 3)$ (f) $(x - 10)(x + 10)$
   (g) $(x + 2)(x + 2)$ or $(x + 2)^2$
   (h) $(x + 6)(x - 6)$ (i) $(x + 5)(x - 5)$
   (j) $(x + 1)(x + 9)$ (k) $(x + 9)(x - 1)$
   (l) $(x + 1)(x - 9)$ (m) $(x - 1)(x - 9)$
   (n) $x(x - 5)$

2. (a) $(2x+1)(x-3)$ (b) $(3x+1)(x-2)$
  (c) $(5x+3)(2x+1)$
  (d) $2x^2+12x+16$ has a common factor of 2, which should be written outside a bracket to give $2(x^2+6x+8)$. The bracket can then be factorised to give $2(x+4)(x+2)$.
  (e) $(2x+3)(x+1)$ (f) $(3s+2)(s+1)$
  (g) $(3z+2)(z+5)$
  (h) $9(x^2-4)=9(x+2)(x-2)$
  (i) $(2x+5)(2x-5)$
3. (a) $(x+y)(x-y)=x^2+yx-yx-y^2$
    $=x^2-y^2$
  (b) (i) $(4x+1)(4x-1)$

  (ii) $(4x+3)(4x-3)$
  (iii) $(5t-4r)(5t+4r)$
4. (a) $(x-2)(x+5)$ (b) $(2x+5)(x-4)$
  (c) $(3x+1)(3x-1)$ (d) $10x^2+14x-12$ has a common factor of 2, which is written outside a bracket to give $2(5x^2+7x-6)$. The bracket can then be factorised to give $2(5x-3)(x+2)$; (e) $(x+13)(x+2)$.
  (f) $(-x+1)(x+3)$
5. (a) $(10+7x)(10-7x)$
  (b) $(6x+5y)(6x-5y)$
  (c) $\left(\frac{1}{2}+3v\right)\left(\frac{1}{2}-3v\right)$ (d) $\left(\frac{x}{y}+2\right)\left(\frac{x}{y}-2\right)$

Chapter 9

Self-assessment questions 9.2

1. There are no factors common to both numerator and denominator and hence cancellation is not possible. In particular 3 is not a factor of the denominator.
2. Same as Q1 – there are no factors common to numerator and denominator. In particular x is a factor neither of the denominator nor of the numerator.
3. $\frac{x+1}{2x+2}=\frac{x+1}{2(x+1)}=\frac{1}{2}$

  The common factor, $x+1$, has been cancelled.

Exercise 9.2

1. (a) $\frac{3x}{y}$ (b) $\frac{9}{x}$ (c) $3y$ (d) $3x$ (e) 9
  (f) 3
2. (a) $\frac{5x}{y}$ (b) $\frac{3x}{y}$ (c) $15y$ (d) 15 (e) $-x^2$

(f) $-\frac{1}{y^4}=-y^{-4}$ (g) $\frac{1}{y}=y^{-1}$ (h) y^{-7}

3. (a) $\frac{1}{3+2x}$ (b) $1+2x$ (c) $\frac{1}{2+7x}$
  (d) $\frac{x}{2+7x}$ (e) $\frac{x}{1+7x}$ (f) $\frac{1}{7x+y}$
  (g) $\frac{y}{7x+y}$ (h) $\frac{x}{7x+y}$
4. (a) $5x+1$ (b) $\frac{3(5x+1)}{3(x+2y)}=\frac{5x+1}{x+2y}$
  (c) $\frac{3}{x+2}$ (d) $\frac{3}{y+2}$ (e) $\frac{13}{x+5}$
  (f) $\frac{17}{9y+4}$
5. (a) $\frac{1}{3+2x}$ (b) $\frac{2}{x+7}$
  (c) $\frac{2(x+4)}{(x-2)(x+4)}=\frac{2}{x-2}$ (d) $\frac{7}{ab+9}$
  (e) $\frac{y}{y+1}$

6. (a) $\dfrac{1}{x-2}$ (b) $\dfrac{2(x-2)}{(x-2)(x+3)} = \dfrac{2}{x+3}$

(c) $\dfrac{1}{x+2}$ (d) $\dfrac{(x+1)(x+1)}{(x+1)(x-3)} = \dfrac{x+1}{x-3}$

(e) $\dfrac{2}{x-3}$ (f) $\dfrac{1}{x-3}$ (g) $\dfrac{1}{2(x-3)}$

(h) $\dfrac{2}{x-3}$ (i) $\dfrac{1}{2(x+4)}$ (j) $\dfrac{1}{2}$ (k) 2

(l) $x+4$ (m) $\dfrac{1}{x-3}$ (n) $\dfrac{1}{x+4}$ (o) $\dfrac{1}{2}$

(p) $\dfrac{x+4}{2x+9}$

Self-assessment question 9.3

1. True

Exercise 9.3

1. (a) $\dfrac{y}{6}$ (b) $\dfrac{z}{6}$ (c) $\dfrac{2}{5y}$ (d) $\dfrac{2}{5x}$ (e) $\dfrac{3x}{4y}$

(f) $\dfrac{3x^2}{5y}$ (g) $\dfrac{3x}{5y}$ (h) $\dfrac{7x}{16y}$ (i) $\dfrac{1}{4x}$ (j) $\dfrac{x}{4}$

(k) $\dfrac{1}{x}$ (l) $\dfrac{x}{9}$ (m) $\dfrac{1}{x}$ (n) $\dfrac{1}{9x}$ (o) $\dfrac{x}{2}$

2. (a) $\dfrac{1}{x}$ (b) $\dfrac{x}{4}$ (c) 1 (d) x (e) $\dfrac{1}{x}$

(f) $\dfrac{4}{x}$ (g) $\dfrac{6}{x}$

3. (a) $\dfrac{a}{20}$ (b) $\dfrac{5a}{4b}$ (c) $\dfrac{1}{2ab}$ (d) $\dfrac{6x^2}{y^3}$

(e) $\dfrac{3b}{5a^2}$ (f) $\dfrac{x}{4y}$ (g) $\dfrac{x}{3(x+y)}$ (h) $\dfrac{1}{3(x+4)}$

4. (a) $\dfrac{3y}{z^3}$ (b) $\dfrac{3+x}{y}$ (c) $\dfrac{x}{12}$ (d) $\dfrac{b}{c}$

5. (a) $\dfrac{1}{x+4}$ (b) $\dfrac{4(x-2)}{x}$

(c) $\dfrac{12ab}{5ef} \times \dfrac{f}{4ab^2} = \dfrac{3}{5eb}$

(d) $\dfrac{x+3y}{2x} \times \dfrac{4x^2}{y} = \dfrac{2x(x+3y)}{y}$ (e) $\dfrac{9}{xyz}$

6. $\dfrac{2}{x+3}$

7. $\dfrac{x+4}{x+3}$

Self-assessment question 9.4

1. The l.c.d. is the simplest expression that is divisible by all the given denominators. To find the l.c.d. first factorise all denominators. The l.c.d. is then formed by including the minimum number of factors from the denominators such that each denominator can divide into the l.c.d.

Exercise 9.4

1. (a) $\dfrac{5z}{6}$ (b) $\dfrac{7x}{12}$ (c) $\dfrac{6y}{25}$

2. (a) $\dfrac{x+2}{2x}$ (b) $\dfrac{1+2x}{2}$ (c) $\dfrac{1+3y}{3}$

(d) $\dfrac{y+3}{3y}$ (e) $\dfrac{8y+1}{y}$

3. (a) $\dfrac{10-x}{2x}$ (b) $\dfrac{5+2x}{x}$ (c) $\dfrac{9-x}{3x}$

(d) $\dfrac{2x-3}{6}$ (e) $\dfrac{9+x}{3x}$

4. (a) $\dfrac{3}{x}+\dfrac{4}{y} = \dfrac{3y}{xy}+\dfrac{4x}{xy} = \dfrac{3y+4x}{xy}$

(b) $\dfrac{3}{x^2}+\dfrac{4y}{x} = \dfrac{3}{x^2}+\dfrac{4xy}{x^2} = \dfrac{3+4xy}{x^2}$

(c) $\dfrac{4ab}{x}+\dfrac{3ab}{2y} = \dfrac{8aby}{2xy}+\dfrac{3abx}{2xy} = \dfrac{8aby+3abx}{2xy}$

(d) $\dfrac{4xy}{a}+\dfrac{3xy}{2b} = \dfrac{8xyb}{2ab}+\dfrac{3xya}{2ab}$

$= \dfrac{8xyb+3xya}{2ab}$

(e) $\dfrac{3}{x} - \dfrac{6}{2x} = \dfrac{3}{x} - \dfrac{3}{x} = 0$

(f) $\dfrac{3x}{2y} - \dfrac{7y}{4x} = \dfrac{6x^2}{4xy} - \dfrac{7y^2}{4xy} = \dfrac{6x^2 - 7y^2}{4xy}$

(g) $\dfrac{3}{x+y} - \dfrac{2}{y} = \dfrac{3y}{(x+y)y} - \dfrac{2(x+y)}{(x+y)y}$

$= \dfrac{3y - 2x - 2y}{(x+y)y} = \dfrac{y - 2x}{(x+y)y}$

(h) $\dfrac{1}{a+b} - \dfrac{1}{a-b}$

$= \dfrac{a-b}{(a+b)(a-b)} - \dfrac{a+b}{(a+b)(a-b)}$

$= \dfrac{a-b-a-b}{(a+b)(a-b)} = \dfrac{-2b}{(a+b)(a-b)}$

(i) $2x + \dfrac{1}{2x} = \dfrac{4x^2}{2x} + \dfrac{1}{2x} = \dfrac{4x^2 + 1}{2x}$

(j) $2x - \dfrac{1}{2x} = \dfrac{4x^2}{2x} - \dfrac{1}{2x} = \dfrac{4x^2 - 1}{2x}$

5. (a) $\dfrac{x}{y} + \dfrac{3x^2}{z} = \dfrac{xz}{yz} + \dfrac{3x^2y}{yz} = \dfrac{xz + 3x^2y}{yz}$

(b) $\dfrac{4}{a} + \dfrac{5}{b} = \dfrac{4b + 5a}{ab}$

(c) $\dfrac{6x}{y} - \dfrac{2y}{x} = \dfrac{6x^2 - 2y^2}{xy}$

(d) $3x - \dfrac{3x+1}{4} = \dfrac{12x}{4} - \dfrac{3x+1}{4}$

$= \dfrac{12x - 3x - 1}{4} = \dfrac{9x - 1}{4}$

(e) $\dfrac{5a}{12} + \dfrac{9a}{18} = \dfrac{15a + 18a}{36} = \dfrac{33a}{36} = \dfrac{11a}{12}$

(f) $\dfrac{x-3}{4} + \dfrac{3}{5} = \dfrac{5(x-3) + 12}{20}$

$= \dfrac{5x - 15 + 12}{20} = \dfrac{5x - 3}{20}$

6. (a) $\dfrac{2x+3}{(x+1)(x+2)}$ (b) $\dfrac{3x+1}{(x-1)(x+3)}$

(c) $\dfrac{4x+17}{(x+5)(x+4)}$ (d) $\dfrac{4x-10}{(x-2)(x-4)}$

(e) $\dfrac{5x+4}{(2x+1)(x+1)}$ (f) $\dfrac{x+1}{x(1-2x)}$

(g) $\dfrac{3x+7}{(x+1)^2}$ (h) $\dfrac{x}{(x-1)^2}$

Exercise 9.5

1. (a) $\dfrac{4}{x+2} + \dfrac{3}{x+3}$ (b) $\dfrac{5}{x+4} - \dfrac{3}{x+1}$

(c) $\dfrac{1}{x+6} - \dfrac{2}{2x+3}$ (d) $\dfrac{2}{x-1} + \dfrac{3}{x-4}$

(e) $\dfrac{5}{2x-1} - \dfrac{1}{x+2}$

2. $\dfrac{5}{2(3x-2)} - \dfrac{3}{2(2x+3)}$

3. (a) $\dfrac{4}{x+5} - \dfrac{3}{x-5}$ (b) $\dfrac{1}{x-3} - \dfrac{1}{(x-3)^2}$

(c) $\dfrac{3}{1-x} - \dfrac{2}{x+2}$ (d) $\dfrac{4}{3x-1} - \dfrac{1}{(3x-1)^2}$

Chapter 15

Self-assessment questions 15.1

1. $\log A + \log B = \log AB$

$\log A - \log B = \log\left(\dfrac{A}{B}\right)$

$\log A^n = n \log A$

Exercise 15.1

1. (a) $\log 5 + \log 9 = \log(5 \times 9) = \log 45$
   (b) $\log 9 - \log 5 = \log(\tfrac{9}{5}) = \log 1.8$
   (c) $\log 5 - \log 9 = \log(\tfrac{5}{9})$
   (d) $2 \log 5 + \log 1 = 2 \log 5 = \log 5^2$
   $= \log 25$

(e) $2 \log 4 - 3 \log 2 = \log 4^2 - \log 2^3$
$= \log 16 - \log 8 = \log(\frac{16}{8}) = \log 2$

(f) $\log 64 - 2 \log 2 = \log(64/2^2) = \log 16$

(g) $3 \log 4 + 2 \log 1 + \log 27 - 3 \log 12$
$= \log 4^3 + 2(0) + \log 27 - \log 12^3$
$= \log \left(\dfrac{4^3 \times 27}{12^3} \right)$
$= \log 1 = 0$

2. (a) $\log 3x$ (b) $\log 8x$ (c) $\log 1.5$
   (d) $\log T^2$ (e) $\log 10X^2$

3. (a) $3 \log X - \log X^2 = 3 \log X - 2 \log X$
   $= \log X$

   (b) $\log y - 2 \log \sqrt{y} = \log y - \log(\sqrt{y})^2$
   $= \log y - \log y = 0$

   (c) $5 \log x^2 + 3 \log \dfrac{1}{x} = 10 \log x - 3 \log x$
   $= 7 \log x$

   (d) $4 \log X - 3 \log X^2 + \log X^3$
   $= 4 \log X - 6 \log X + 3 \log X$
   $= \log X$

   (e) $3 \log y^{1.4} + 2 \log y^{0.4} - \log y^{1.2}$
   $= 4.2 \log y + 0.8 \log y - 1.2 \log y$
   $= 3.8 \log y$

4. (a) $\log 4x - \log x = \log(4x/x) = \log 4$
   (b) $\log t^3 + \log t^4 = \log(t^3 \times t^4) = \log t^7$

   (c) $\log 2t - \log \left(\dfrac{t}{4} \right) = \log \left(2t \div \dfrac{t}{4} \right)$
   $= \log \left(2t \times \dfrac{4}{t} \right) = \log 8$

   (d) $\log 2 + \log \left(\dfrac{3}{x} \right) - \log \left(\dfrac{x}{2} \right)$
   $= \log \left(2 \times \dfrac{3}{x} \right) - \log \left(\dfrac{x}{2} \right)$
   $= \log \left(\dfrac{6}{x} \div \dfrac{x}{2} \right)$

$= \log \left(\dfrac{6}{x} \times \dfrac{2}{x} \right)$
$= \log \left(\dfrac{12}{x^2} \right)$

(e) $\log \left(\dfrac{t^2}{3} \right) + \log \left(\dfrac{6}{t} \right) - \log \left(\dfrac{1}{t} \right)$
$= \log \left(\dfrac{t^2}{3} \times \dfrac{6}{t} \right) - \log \left(\dfrac{1}{t} \right)$
$= \log(2t) - \log \left(\dfrac{1}{t} \right)$
$= \log \left(\dfrac{2t}{1/t} \right)$
$= \log 2t^2$

(f) $2 \log y - \log y^2 = \log y^2 - \log y^2 = 0$

(g) $3 \log \left(\dfrac{1}{t} \right) + \log t^2 = \log \left(\dfrac{1}{t} \right)^3 + \log t^2$
$= \log \left(\dfrac{1}{t^3} \right) + \log t^2$
$= \log \left(\dfrac{t^2}{t^3} \right)$
$= \log \left(\dfrac{1}{t} \right) = -\log t$

(h) $4 \log \sqrt{x} + 2 \log \left(\dfrac{1}{x} \right)$
$= 4 \log x^{0.5} + \log \left(\dfrac{1}{x} \right)^2$

$$= \log x^2 + \log \left(\frac{1}{x^2} \right)$$

$$= \log \left(x^2 \frac{1}{x^2} \right)$$

$$= \log 1 = 0$$

(i) $2 \log x + 3 \log t = \log x^2 + \log t^3$
$$= \log(x^2 t^3)$$

(j) $\log A - \dfrac{1}{2} \log 4A = \log A - \log(4A)^{1/2}$

$$= \log \left[\frac{A}{(4A)^{1/2}} \right]$$

$$= \log \left(\frac{A^{1/2}}{2} \right)$$

(k) $\dfrac{\log 9x + \log 3x^2}{3} = \dfrac{\log(9x.3x^2)}{3}$

$$= \frac{\log(27x^3)}{3}$$

$$= \frac{\log(3x)^3}{3}$$

$$= \frac{3 \log(3x)}{3}$$

$$= \log(3x)$$

(l) $\log xy + 2 \log \left(\dfrac{x}{y} \right) + 3 \log \left(\dfrac{y}{x} \right)$

$$= \log xy + \log \left(\frac{x^2}{y^2} \right) + \log \left(\frac{y^3}{x^3} \right)$$

$$= \log \left(xy \frac{x^2}{y^2} \frac{y^3}{x^3} \right)$$

$$= \log \left(\frac{x^3 y^4}{x^3 y^2} \right)$$

$$= \log y^2$$

(m) $\log \left(\dfrac{A}{B} \right) - \log \left(\dfrac{B}{A} \right) = \log \left(\dfrac{A}{B} \div \dfrac{B}{A} \right)$

$$= \log \left(\frac{A}{B} \times \frac{A}{B} \right) = \log \left(\frac{A^2}{B^2} \right)$$

(n) $\log \left(\dfrac{2t}{3} \right) + \dfrac{1}{2} \log 9t - \log \left(\dfrac{1}{t} \right)$

$$= \log \left(\frac{2t}{3} \right) + \log(9t)^{1/2} - \log t^{-1}$$

$$= \log \left[\frac{2t}{3} (9t)^{1/2} \right] + \log t$$

$$= \log \left[\frac{2t}{3} (9t)^{1/2} t \right]$$

$$= \log \left(\frac{2t}{3} 3t^{1/2} t \right)$$

$$= \log(2t^{2.5})$$

5. $\log_{10} X + \ln X = \log_{10} X + \dfrac{\log_{10} X}{\log_{10} e}$

$$= \log_{10} X + 2.3026 \log_{10} X$$

$$= 3.3026 \log_{10} X$$

$$= \log X^{3.3026}$$

6. (a) $\log(9x - 3) - \log(3x - 1)$

$$= \log \left(\frac{9x - 3}{3x - 1} \right) = \log 3$$

(b) $\log(x^2 - 1) - \log(x + 1)$

$$= \log \left(\frac{x^2 - 1}{x + 1} \right) = \log(x - 1)$$

(c) $\log(x^2 + 3x) - \log(x + 3)$

$$= \log\left(\frac{x^2 + 3x}{x + 3}\right) = \log x$$

Chapter 17

Self-assessment questions 17.1

1.   gradient $= m =$

$$\frac{\text{difference in } y \text{ (vertical) coordinates}}{\text{difference in } x \text{ (horizontal) coordinates}}$$

$$m = \frac{y_2 - y_1}{x_2 - x_1}$$

2. The value of c is given by the intersection of the straight line graph with the vertical axis.

Exercise 17.1

1. $y = 2x + 5$

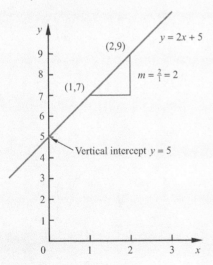

2. $y = 6x - 10$
3. $y = 2$
4. $y = x$
5. $y = -x$
6. $y = 2x + 8$
7. (b), (c) and (d) lie on the line.
8. The equation of the line is $y = mx + c$. The gradient is $m = -1$ and so

$$y = -x + c$$

   When $x = -3$, $y = 7$ and so $c = 4$. Hence the required equation is $y = -x + 4$.
9. The gradient of $y = 3x + 17$ is 3. Let the required equation be $y = mx + c$. Since the lines are parallel then $m = 3$ and so $y = 3x + c$. When $x = -1$, $y = -6$ and so $c = -3$. Hence the equation is $y = 3x - 3$.
10. The equation is $y = mx + c$. The vertical intercept is $c = -2$ and so

$$y = mx - 2$$

   When $x = 3$, $y = 10$ and so $m = 4$. Hence the required equation is $y = 4x - 2$.

Self-assessment questions 17.2

1. A tangent is a straight line that touches a curve at a single point.
2. The gradient of a curve at a particular point is the gradient of the tangent at that point.

Exercise 17.2

1.

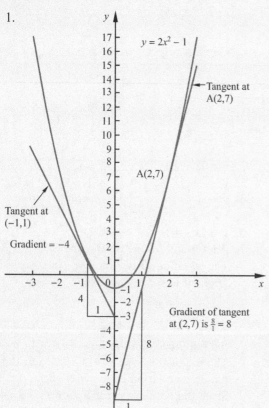

$y = 2x^2 - 1$

Tangent at A(2,7)

A(2,7)

Tangent at (−1,1)

Gradient = −4

Gradient of tangent at (2,7) is $\frac{8}{1} = 8$

2.

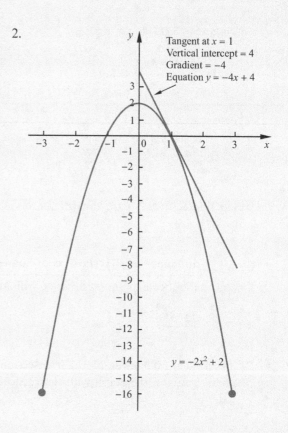

Tangent at $x = 1$
Vertical intercept = 4
Gradient = −4
Equation $y = -4x + 4$

$y = -2x^2 + 2$

Solutions to test and assignment exercises

Chapter 1. Arithmetic of whole numbers

1. a) 4   b) 2   c) 13   d) 4   e) 7   f) 9   g) 6   h) 4   i) 2   j) 3   k) 7   l) 6   m) 6   n) 3   o) 2

2. a) $2 \times 2 \times 2 \times 7$ b) 3×13 c) 2×37

3. a) 2   b) 6   c) 2   d) 8

4. a) 30   b) 143   c) 112   d) 120

Chapter 2. Fractions

1. a) $\frac{11}{12}$ b) $\frac{11}{10}$ c) $\frac{1}{21}$ d) $\frac{43}{30}$ e) $8\frac{5}{12}$ f) $\frac{23}{24}$ g) $\frac{5}{12}$ h) $\frac{23}{60}$

2. a) $\frac{3}{8}$ b) $\frac{4}{9}$ c) $\frac{20}{77}$ d) $\frac{2}{7}$ e) $\frac{13}{8}$ f) $\frac{2}{3}$ g) $-\frac{16}{17}$ h) $\frac{3}{5}$

3. a) $\frac{7}{6}$ b) $\frac{3}{4}$ c) 2   d) $\frac{199}{240}$ e) $\frac{4}{5}$ f) $\frac{5}{4}$ g) $\frac{4}{3}$ h) $\frac{19}{30}$

4. a) $\frac{1}{4}$ b) $\frac{3}{40}$ c) $\frac{2}{5}$ d) $\frac{2}{13}$ e) $\frac{3}{11}$

Chapter 3. Decimal fractions

1. a) $\frac{37}{50}$ b) $\frac{24}{25}$ c) $\frac{1}{20}$ d) $\frac{1}{4}$

2. a) $2\frac{1}{2}$ b) $3\frac{1}{4}$ c) $3\frac{1}{8}$ d) $6\frac{7}{8}$

3. a) 0.317   b) 0.095   c) 0.904

4. a) 0.1   b) 0.098   c) 0.1

5. a) 9.51   b) 9.5   c) 10

6. a) 20.0   b) 20   c) 20

Chapter 4. Percentage and ratio

1. a) 0.08   b) 0.18   c) 0.65

2. a) 37.5%   b) 79%   c) 47.46% (2 d.p.)

3. 408.408

4. 156.25

5. 0.2125

6. $\frac{1}{15}$, $\frac{2}{15}$, $\frac{3}{10}$

7. £ 41.04

8. £ 4600 and £ 8050

9. 8.974,   23.333,   37.692

10. 20% and 30%

11. -10.26% 12. £84   13. -11.03% 14. 12.38% 15. $Z = 0.96X$

Chapter 5. Algebra

1. $44^3 = 85184$, $0.44^2 = 0.1936$, $32.5^3 = 34328.125$.

2. a) $xxxyyyyy = x^3y^5$.   b) $\frac{xxx}{yyyy} = \frac{x^3}{y^4}$,   c) $a^2baab = a^4b^2$.

3. If $x = 2$, $y = 5$, $z = 3$, then $4x^3yz^2 = 1440$.

4. 3.142cm.

5. a) $21^2 - 16^2 = 185$,   b) $(21 - 16)^2 = 25$. In general, $a^2 - b^2 \neq (a - b)^2$.

6. When $x = 4$ and $y = -3$, then

a) $xy = -12$,   b) $\frac{x}{y} = -\frac{4}{3} = -1.3333$,   c) $\frac{x^2}{y^2} = \frac{16}{9} = 1.778$,   d) $\left(\frac{x}{y}\right)^2 = \frac{16}{9} = 1.778$.

7. When $x = 7$, $2x(x + 4) = 14(11) = 154$.

8. When $x = 9$, $4x^2 + 7x = 387$.

9. When $x = -2$, $3x^2 - 7x + 12 = 38$.

10. When $x = -3$, $-x^2 - 11x + 1 = 25$.

11. $I = \frac{V}{R} = \frac{10}{0.01} = 1000$.

12. a) $A = 1$,   b) $A = 1/2$,   c) $A = 1/3$.

13. a) $y = \frac{1}{2}$,   b) y is not defined,   c) $y = 1/12$.

14. a) When n is even, $(-1)^n = 1$. b) When n is odd $(-1)^n = -1$.

15. a) When n is even, $(-1)^{n+1} = -1$. b) When n is odd, $(-1)^{n+1} = 1$.

Chapter 6. Indices

1. a) $\frac{z^5}{z^{-5}} = z^{10}$,   b) $z^0 = 1$,   c) $\frac{z^8z^6}{z^{14}} = 1$.

2. a) $0.25^{1/2} = \pm0.5$ b) $4096^{1/3} = 16$ c) $2601^{1/2} = \pm51$ d) $16^{-1/2} = \pm0.25$.

3. x^8.

4. 1

5. a) $(abc^2)^2 = a^2b^2c^4$ b) $(xy^2z^3)^2 = x^2y^4z^6$ c) $(8x^2)^{-3} = 8^{-3}x^{-6} = \frac{1}{512x^6}$

6. a) $5792 = 5.792 \times 10^3$ b) $98.4 = 9.84 \times 10^1$ c) $0.001 = 1 \times 10^{-3}$ d) $-66.667 = -6.6667 \times 10^1$.

Chapter 7. Simplifying algebraic expressions

1. a) $11x^2 + x$ b) $8 - 17y$ c) $a^2 + a^3 - 2b^2$.
2. a) $-3a^5b^3c$ b) $-x$.
3. a) $7a^2 + 19ab - 6b^2$ b) $x^3 + 2x^2y$ c) $x^3 - xy^2$.
4. a) $21x^2 - x - 2$ b) $-x^2 - 2x + 3$ c) $5x + x^2$ d) $56x^2 + 12x - 8$.
5. a) $6x - 4x^2$ b) $-2a^2 - 5ab - 3b^2$ c) $17x + 2$ d) $-5a + 31$ e) $2b + 2a$.

Chapter 8. Factorization

1. a) $7(x + 7)$ b) $11(11x + 2y)$ c) $a(a + b)$ d) $b(a + b)$ e) $ab(b + a)$.
2. a) $(3x - 2)(x + 1)$ b) $(x + 12)(x - 12)$ c) $(s - 3)(s - 2)$ d) $(2y + 5)(y - 3)$.
3. a) $(1 + x)(1 - x)$ b) $(x + 1)(x - 1)$ c) $(3 + x)(3 - x)$ d) $(x + 9)(x - 9)$ e) $(5 + y)(5 - y)$.
4. a) $x^2 + 6x = x(x + 6)$ b) $s^2 + 3s + 2 = (s + 1)(s + 2)$ c) $s^2 + s - 2 = (s + 2)(s - 1)$
   d) $x^2 + 11x + 28 = (x + 7)(x + 4)$ e) $2x^2 - 17x - 9 = (2x + 1)(x - 9)$.

Chapter 9. Algebraic fractions

1. a) $\dfrac{5}{4 + 3b}$ b) $\dfrac{a}{6a + 3b}$ c) $\dfrac{ab}{6a + 3b}$
   d) $\dfrac{5}{8}$ e) $\dfrac{1}{13 + y}$ f) $13 + y$.

2. a) $5ab$ b) $\dfrac{16}{3x}$ c) $\dfrac{3t - 4}{2}$
   d) $\dfrac{8x + 41}{3}$ e) $\dfrac{61}{6x}$ f) $\dfrac{xy + 3x}{y}$
   g) $\dfrac{x^2y^2 + 1}{xy}$ h) $\dfrac{3x + y}{(x + y)(x - y)}$.

3. a) $\dfrac{25y}{63}$ b) $\dfrac{32}{15x}$ c) $\dfrac{9x^2 + 5y^2}{6xy}$ d) $\dfrac{3m + n}{2}$ e) $\dfrac{m - n}{2}$ f) $\dfrac{m + n}{2}$ g) $\dfrac{5s - 9}{30}$.

4. x.

5. a) $\dfrac{3}{x + 7} + \dfrac{1}{x + 2}$ b) $\dfrac{3}{x + 1} - \dfrac{2}{x + 2}$ c) $\dfrac{2}{x + 4} - \dfrac{1}{x - 5}$ d) $\dfrac{1}{3x + 1} - \dfrac{1}{2x + 1}$ e) $\dfrac{4}{2x + 3} - \dfrac{3}{3x - 1}$

6. a) $\dfrac{5}{x + 3} + \dfrac{3}{x + 2}$ b) $\dfrac{4}{x + 5} - \dfrac{1}{x + 4}$ c) $\dfrac{3}{x + 2} + \dfrac{1}{(x + 2)^2}$
   d) $\dfrac{1}{x - 3} - \dfrac{3}{(x - 3)^2}$ e) $\dfrac{4}{2x - 1} - \dfrac{3}{(2x - 1)^2}$

Chapter 17. The straight line

1. a), b), d) and e) have straight line graphs.

    a) gradient$=2$, intercept$=-11$

    b) gradient$=5$, intercept$=10$

    d) gradient$=3$, intercept$=-3$

    e) gradient$=1$, intercept$=3/2$

2. $y = 7x + 4$.

3. $y = -2x + 3$.

4. $y = 5$. The line does not pass through $(2, 6)$.

5. The graph is shown below.

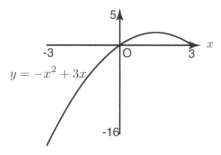

Graph for Q5

Gradient of tangent at $x = -2$ is 7. Gradient of tangent at $x = 1$ is 1.

6. a) $y = -2x$,   b) $y = -4x$,   c) $y = 4x$.

7. $y = 6x - 14$

8. $y = -6x + 3$

9. a) $x = -16/3$, b) $y = 32/5$

10. $y = \frac{3}{2}x + 3$

11. a) $y = -1$, b) $x = \frac{1}{4}$, c) at the point $(\frac{3}{4}, 2)$

Answers to questions that test your understanding

Chapter 10   Transposing formulae

1a) $s = \dfrac{t}{m}$

1b) $s = \dfrac{t+p}{m}$

1c) $s = \dfrac{t}{2} - m$

1d) $s = \dfrac{t}{m} + p$

2a) $b = \dfrac{g}{a}$

2b) $b = g - ac$

2c) $b = \dfrac{c+a}{g}$

2d) $b = \dfrac{gc}{2} + 1$

3a) $m = \sqrt{s}$

3b) $m = s^2 - f$

3c) $m = (2f - s)^2$

3d) $m = \sqrt{\dfrac{f+2}{s}}$

4a) $w = dx^2 - 2$

4b) $w = \sqrt[3]{d - p}$

4c) $w = \dfrac{d^3}{p}$

4d) $w = \sqrt[3]{\dfrac{2d}{p}}$

5a) mass = moles × molar mass

5b) mass = molarity × molar mass × volume

6) $c = \dfrac{A}{\alpha l}$

7) $r = \sqrt[3]{\dfrac{3V}{4\pi}}$

8a) $V_{max} = \dfrac{v(K_m + [S])}{[S]}$

8b) $K_m = \dfrac{V_{max}[S]}{v} - [S]$

Chapter 11 Measurement

1a) 1.2 mm

1b) 4 mm

1c) 3.5 mm

2a) 45 g

2b) 0.036 g

2c) 0.064 g

3a) 0.0035 m^2

3b) 0.00000048 m^2 or 4.8 x 10^{-7} m^2

3c) 500 m^2

4a) 4000 mm^3

4b) 0.000855 mm^2 or 8.53 x 10^{-4} mm^3

4c) 4 mm^3

5a) 34 ml

5b) 5000 ml

5c) 0.68 ml

6a) 43 litres

6b) 0.3 litres

6c) 4.6 litres

7a) 0.93 μm^2

7b) 7.47 x 10^{-2} μm^2 or 0.0747 μm^2

7c) 4.71 x 10^2 μm^2 or 471 μm^2

8a) 3.51 x 10^2 km^2 or 351 km^2

8b) 8.96 x 10^2 km^2 or 896 km^2

8c) 0.764 km^2

9a) 7.92 mm^3

9b) 8.335 x 10^{-3} mm^3

9c) 2 x 10^{-2} mm^3 or 0.02 mm^3

10a) 4.3×10^{-5} km^3 or 0.000043 km^3

10b) 9.46×10^{-2} km^3 or 0.0946 km^3

10c) 3.7×10^{-7} km^3

11a) 17.64 mm^2

11b) 0.314 m^2

11c) 79.51 cm^3

11d) 154 mm^2

11e) 6.61 m^2

11f) 0.205 μm^3 or 2.05×10^{-4} μm^3

11g) 796 m^3

11h) 0.00021 m^3

11i) 65.5 km^3

11j) 1539 μm^3

11k) 7.19 m^3

11l) 36.1 cm^3

12a) 108 acorns

12b) 83.5 km

12c) 12.5 g l^{-1}

12d) 8,000 trees

12e) 1.2×10^{10} cod

12f) 4×10^8 woodlice

12g) 1.92×10^{11} bacteria

12h) 9.75×10^7 grass plants

12i) 6.78×10^8 cells

12j) 24 litres

12k) 0.347 eggs s^{-1}

12l) 8.4×10^7 bluebells

12m) 1.2×10^8 *Amoeba*

12n) 6.2 m^2

12o) 23,000 litres

13a) 905 μm^3

13b) 489 mm^3

13c) 551mm^2

13d) Top = 82.3 cm^3

Stalk = 6.44 cm^3

Biomass = 13.3 g

13e) Volume = 17159.5 μm^3

Mass = 0.0515 μg

13f) Volume of bacterium = 0.0718 μm^3

Volume of colony = 3.595×10^{10} μm^3

13g) Volume of worm = 4450 mm^3

Mass of worm = 1.78 g

13h) Volume of one worm = 356.4 mm^3

Total volume of worms = 1.14 m^3

NB. The answers given below should be open for discussion in terms of the number of significant figures and decimal places provided.

14a) 0.4 m s^{-1}

14b) 10.76 m^2

14c) 17.9 cm^3

14d) 938 cells

14e) Area = 56.21 cm^2, circumference = 26.58 cm, volume = 301.29 cm^3

14f) 6.5 m s^{-1}

14g) After 5 months = 4.5 m

After 9 years = 15.7 m

14h) 2.8 cm^2 day^{-1}

14i) 22.8 kg m^{-2}

14j) 4.100 litres, 4100 ml or 4099.6 ml

14k) 0.756 m

14l) 3285 strides

14m) 5.3×10^4 cells ml^{-1}

Chapter 12 Dilution and Scaling

Percentage and Ratio

1a) 168

1b) 3:5

1c) 8:5.76:18.24

1d) 3:1:1

1e) 691 plants

2a) 2600 patients

2b) 10.4 %

2c) severe = 1.26 %, mild = 5.88 %, no symptoms = 92.86 %

2d) 15.38 %

2e) 0.48

2f) 0.62

Dilution

3a) 24 mg ml^{-1}

3b) 0.24

3c) dilute by 20 (or make a 1/20 dilution)

3d) 1.5 x 10^4 cell ml^{-1}, 7.5 x 10^3 cells ml^{-1}, 5 x 10^3 cells ml^{-1}

3e) 2 mg ml^{-1}

3f) 1/5

3g) 6.67 x 10^2 bacteria ml^{-1}, 16.6 %, 6:1

Scaling

4a) 12 μm

4b) 5 x 10^{-4} mm or 0.5 μm

4c) 0.34 mm

4d) 100000:1

4e) 1.5 km

4f) estimating that the eye piece scale = 0.33 mm: Specimen = 0.297 mm or 297 μm

5a) 2.5 x 10^8 cfu g^{-1}

5b) 8 x 10^6 cfu (100 ml)$^{-1}$

Chapter 13 Trigonometry

1a) 7.2 mm

1b) 11.69 cm

1c) 32 m

1d) 1.6 μm

2a) 17.5°

2b) 54.9°

2c) 61.4°

2d) 70.9°

3a) 64.3 °

3b) 36.9 °

3c) 33.2 °

3d) 33.3 °

4a) 47 cm

4b) 0.85 μm

4c) 259 mm

4d) 92.4 m

5a) 70 μm

5b) 4.21 μm

5c) 44 m

5d) 49.3 cm

6a) 2.13 m

6b) 48.8 °

Chapter 14   Moles and Molarity

1a) 0.048 moles

1b) 56 g mol^{-1}

1c) 0.17 mol l^{-1}

1d) 0.377 mol l^{-1}

1e) 100 ml

1f) NaCl = 4 ml, KOH = 4.5 ml

1g) 166 g mol^{-1}

1h) 2 litres

2a) 7.87 x 10^{-5} mol

2b) 1.43 x 10^{-5} mol l^{-1}

2c) 0.015 ml

2d) 1.94 x 10^{-8} mol

2e) 0.093 g

2f) 2.67 x 10^{-4} mol l^{-1}

2g) 4.15 x 10^{-5} g

2h) 0.51 ml

3a) 2.05 x 10^{-3} mol l^{-1}

3b) 5.48x10^{-4} g

3c) 0.0275 μmol l^{-1} below

Chapter 15 Logarithms

1a) $\log_2 16 = 4$

1b) $\log_8 512 = 3$

1c) $\log_4 64 = 3$

1d) $\log_3 243 = 5$

1e) $\log_5 25 = 2$

2a) $2^3 = 8$

2b) $4^3 = 64$

2c) $7^4 = 2401$

2d) $9^2 = 81$

2e) $3^5 = 243$

3a) $\log_2 4 = 2$

3b) $4^4 = 16$

3c) $\log_4 64 = 3$

3d) $10^5 = 100,000$

3e) $3^3 = 27$

4a) $\log 24\,x^3$

4b) $\log 2\,x^{-1}$

4c) $\log \frac{1}{2xy}$

4d) $\log 6t^3$

4e) $\log 14$

4f) $\log 2$

4g) $\log \frac{t^2}{16}$

5a) pH 4.4

5b) pH 5.2

5c) moles acetic acid $= \frac{200}{1000} \times 0.1 = 0.02$

moles Na acetate $= \frac{150}{1000} \times 0.2 = 0.03$

[acetic acid] $= 0.02 \times \frac{1000}{350} = 0.057$ mol l^{-1}

[Na acetate] $= 0.03 \times \frac{1000}{350} = 0.086$ mol l^{-1}

pH $= 4.7 + \log \frac{0.086}{0.057}$ pH $= 4.88$

5d) [acid] $= 0.005 \times \frac{1000}{300} = 0.017$ mol l^{-1}

[base] $= 0.06 \times \frac{1000}{300} = 0.2$ mol l^{-1}

5d) contd.

pH $= 4.7 + \log \frac{0.2}{0.017} = 5.77$ 6a) pH 5

6b) pH 8

6c) pH 3.3

6d) pH 11.8

6e) pH 12.5

6f) pH 6.3

7)

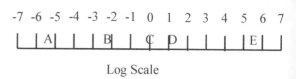

-7 -6 -5 -4 -3 -2 -1 0 1 2 3 4 5 6 7

Log Scale

8) Values to put on the graph:

log volume (µm³)	log Amoeba
1	0.60
9	4.30
11.60	7.78
19.26	12.70

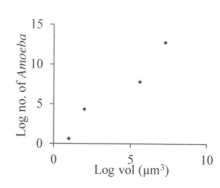

9a) 16.5 years

9b) 8.06 x 10^{-3} yr^{-1}

9c) 347 years

9d) 1.98 days

9e) -0.032 s^{-1}

9f) 3.5 days

9g) -0.17 ms^{-1}

Chapter 16 Graphs

1a)

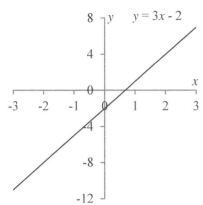

1b)

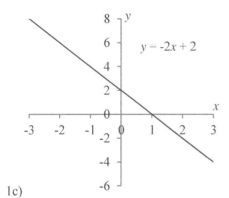

1c)

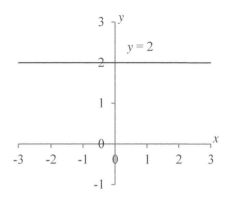

1d)

1e)

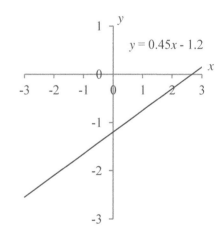

1c)

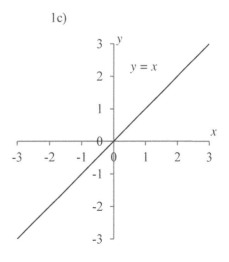

2) Cub b had the faster growth rate.

Cub a had the higher birth mass.

3a) *cl* is plotted on the *x* axis

A is plotted on the *y* axis

3b) *E* would have a value of 3.24

The graph would intercept the *y* axis at 0.

4) 1.3 m min^{-1}

5)

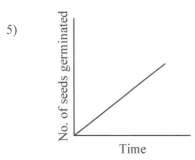

Number of seed germinating is directly proportional to time; as time increases, so more seeds germinate.

6)

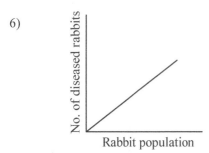

Number of diseased rabbits = increase in disease x number in the population

7)

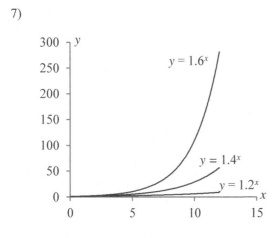

8) $\ln y = 0.053x + \ln 0.683$

9a) *x* would be $\frac{v}{[S]}$ *y* would be *v*

9b) *x* would be *[S]*   *y* would be $\frac{[S]}{v}$

10) Gradient = *k*

11a) a minimum point

11b) 98.53 µg l^{-1}

11c) 56.9 µm d^{-1}

12a) $y = (x + 2)(x + 1)$
12b) $y = (x - 9)(x + 1)$
12c) $y = (x + 6)(x - 6)$
12d) $y = (x + 0)(x - 5)$ or $y = x(x - 5)$
12e) $y = (x + 2)(x + 2)$

13a) $y = x^2 - 25$
13b) $y = x^2 - 100$
13c) $y = x^2 - 11x + 24$

Chapter 17 Calculus

1a) $y = 2.5x + 0.5$

1b) $y = 2.5x + 8$

1c) $y = -3x + 10$

1d) $y = 5$

1e) $y = 0.5x + 3.3$

1f) $y = -0.34x + 6.68$

1g) $y = 18.19x - 5.80$

2a) $\frac{dy}{dx} = 2x$

2b) $\frac{dy}{dx} = -12x^2$

2c) $\frac{dy}{dx} = 6x$

2d) $\frac{dy}{dx} = 2x$

2e) $\frac{dy}{dx} = 6x$

2f) $\frac{dy}{dx} = -4x$

2g) $\frac{dy}{dx} = 4x + 4$

2h) $\dfrac{dy}{dx} = -6x^2 + 3$

2i) $\dfrac{dy}{dx} = 6x - 4$

2j) $\dfrac{dy}{dx} = -x^{-2}$

2k) $\dfrac{dy}{dx} = -4x^{-3}$

2l) $\dfrac{dy}{dx} = 2x^{-2}$

2m) $\dfrac{dy}{dx} = 2e^{2x}$

2n) $\dfrac{dy}{dx} = 2e^{x}$

2o) $\dfrac{dy}{dx} = 12e^{4x}$

3a) $\dfrac{dy}{dt} = 2t$

3b) $\dfrac{dy}{dt} = 4t$

3c) $\dfrac{dy}{dt} = -4t$

3d) $\dfrac{dy}{dt} = 4t - 7$

3e) $\dfrac{dy}{dt} = -9t^2 - 4$

3f) $\dfrac{dy}{dt} = 8t + 7$

3g) $\dfrac{dy}{dt} = -2t^{-3}$

3h) $\dfrac{dy}{dt} = -2t^{-2}$

3i) $\dfrac{dy}{dt} = 6e^{3t}$

5a) 14.48 h
5b) 45.78 nl^{-1}

6a) True
6b) False

4.

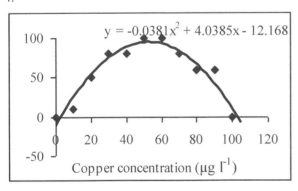

$$y = -0.0381x^2 + 4.0385x - 12.168$$

Copper concentration (μg l^{-1})

Copper concentration = 53 μg l^{-1}

Maximum growth = 94.8 μm d^{-1}